String Theory and the Real World (Second Edition)

The visible sector

String Theory and the Real World (Second Edition)

The visible sector

Gordon Kane

Randall Physics Lab, University of Michigan, Ann Arbor, Michigan, USA

IOP Publishing, Bristol, UK

ISBN   978-0-7503-3583-6 (ebook)
ISBN   978-0-7503-3581-2 (print)
ISBN   978-0-7503-3584-3 (myPrint)
ISBN   978-0-7503-3582-9 (mobi)

DOI   10.1088/978-0-7503-3583-6

Version: 20210701

IOP ebooks

British Library Cataloguing-in-Publication Data: A catalogue record for this book is available from the British Library.

Published by IOP Publishing, wholly owned by The Institute of Physics, London

IOP Publishing, Temple Circus, Temple Way, Bristol, BS1 6HG, UK

US Office: IOP Publishing, Inc., 190 North Independence Mall West, Suite 601, Philadelphia, PA 19106, USA

To my wife, Lois, who still finds my physics life to be a grand adventure.

Some of us should venture to embark on a synthesis of facts, and theories, albeit with second-hand and incomplete knowledge of some of them—and at the risk of making fools of ourselves.

—Erwin Schrödinger, What is Life? (1943)

Scientists of the past were not just like scientists of today who didn't know as much as we do. They had completely different ideas of what there was to know or how you go about learning it.

—Steven Weinberg

Contents

Preface to second edition

Although it is only four years since the first edition was written, there are several reasons why a new edition of this book is appropriate. Partly it is longer because **significant new work in several areas, such as identifying the inflaton (the physical field that generates inflation) and also the mechanism for the matter asymmetry, can now be described**, and partly we understand the whole framework better and can extend explanations. The goals of the publisher have changed to prefer a longer book and I can take advantage of that. Finally, the book is about double in length. It includes a useful glossary. I wish most books were under 200 pages, since I might then find time to read them, and I have kept it shorter than that.

One contribution **I hope people take away from this book is the list of about twenty questions or issues such that a model or theory that answers all of them is a good candidate for a theory which describes and explains our world**—what Steven Weinberg has called a final theory. The world seems messy and full of complicated things. But we don't need to calculate the predictions for hundreds of experiments and observations to think we have a good theory—about twenty is enough. These issues are ones such as what are we made of (quarks and electrons), what is the origin of mass for the fundamental particles, quarks, electrons, etc (their interaction with Higgs bosons), and so on. Most of them will be familiar to most readers, and what matters here is the questions that elucidate each part of the world, not the details or calculations.

Part I should be pretty accessible to any curious reader. Part II is mainly to document the answers to issues such as hidden sectors, inflation, the matter asymmetry, etc. Four short appendices give answers and clarifications.

I'm using capitalization in a way I think is similar to how it was used for the Standard Model. We use lower case for 'final theory' until we're convinced it's correct, and then caps. The final theory proposed here could be correct, but needs to be studied more before we're confident.

Although some topics are somewhat technical, I hope the first edition was about at Scientific American level, and I will try to make this edition even more pedagogical and comprehensible. Every technical term is explained in a good Glossary, so the reader can quickly flip to the Glossary when a new or forgotten term occurs.

The reader may have noticed that I plan to say crucial aspects in bold type, so one can skim the bold statements and have a summary of the results. At first the bold words may occur often, but after the initial ideas and results they will only occur occasionally.

We're not quite at the stage where we can definitively answer all the questions, but in one sense we're close. We can probably answer all but two and we have good candidate answers for the last two. It will of course be crucial to have data on at least one superpartner, and on dark matter, as will be discussed in the book. We can relate the answers to a deeper underlying theory, M-theory in ten space dimensions, compactified to our four-dimensional space–time plus seven space dimensions curled up in a manifold with a size about the Planck scale, and originating from an E_8 singularity. All these words will be defined and explained as we go along. When the first edition was written we could hope the M-theory compactification was good; now for this edition we can see how a complete picture based on M-theory compactification may have emerged. We don't yet know what the dark matter is, nor the superpartner mass spectrum, but we can have a theory that includes them as candidates and is otherwise complete. People who study compactified string theories from the point of view of answering the questions of chapter 6 (with collaborators) include Michele Cicoli, Fernando Quevedo, and Hans Peter Nilles.

In addition to those I acknowledged for the first edition, I am particularly grateful to Martin Winkler for a productive collaboration and his significant contributions, and I appreciate discussions with Gary Shiu, Scott Watson, and my students Eric Gonzalez and Khoa Dang Nguyen.

Preface

Since this book is meant to present a point of view rather than a review, or to be a balanced presentation, it will have a few references to assist finding information, but only a few. Since compactified M-theory is emphasized because it has been my focus, some references to its early history will be included. Otherwise a few references will be given to help anyone who wants to find further information. I apologize to many people who would have been referenced in a review, and I hope they understand the role of references in a book such as this.

I am very grateful to Professor Bobby Acharya, at King's College London, with whom I have collaborated for over a decade in this area, who has taught me a great deal, and without whom the work would not have been done. I also want to thank a number of colleagues for collaboration and/or discussions: Konstantin Bobkov, Sebastian Ellis, Daniel Feldman, Joel Giedt, David Gross, Eric Kuflik, Piyush Kumar, Ran Lu, Brent Nelson, Malcolm Perry, Aaron Pierce, Jing Shao, Kuver Sinha, Diana Vaman, Scott Watson, and Bob Zheng.

Author biography

Gordon Kane

 Gordon Kane got his PhD at University of Illinois. He joined the University of Michigan in 1965, and became Professor of Physics in 1975. He is now Victor Weisskopf Distinguished University Professor of Physics, and Adjunct Professor in the School of Art and Design. He retired from teaching in 2020, and is now Active Emeritus. He was Director of the Michigan Center for Theoretical Physics from 2005 to 2011. In 1971–72 he was a J S Guggenheim Fellow at Rutherford Laboratory and Oxford, in 1986 Scientific Associate at CERN, and in 2007 he was Member, Institute for Advanced Study, Princeton. He has been elected a Fellow of the American Physical Society, Fellow of the American Association for the Advancement of Science, and Fellow of the British Institute for Physics. He was awarded the 2012 Julius Edgar Lilienfeld Prize of the American Physical Society, and co-awarded the 2017 Sakurai Prize of the American Physical Society. He was a 2009 Member of the Triennial Committee of Visitors of the National Science Foundation and Chair of its Theoretical Physics Subpanel, and has served on the High Energy Physics and Scientific Advisory Committees at the Stanford Linear Accelerator Center and at Brookhaven National Laboratory. He originated and co-organized an International Summer School in String Phenomenology, at Simons Center in 2012. Gordon Kane's research has covered several areas of physics beyond the Standard Model(s) of particle physics and cosmology, including supersymmetry, Higgs physics, dark matter, cosmology, collider physics, and string phenomenology. He has published over 200 scientific papers, given over 250 scientific talks, and written or edited ten books. Two are for the general public, in particular *Supersymmetry and Beyond, From the Higgs Boson to the New Physics*, Basic Books 2013. His most recent book, *String Theory and the Real World*, is at Scientific American level.

Part I

What we probably know

IOP Publishing

String Theory and the Real World (Second Edition)
The visible sector
Gordon Kane

Chapter 1

The Standard Models—overview and perspective

My aim in this book is to explain how 'string theory' (actually M-theory) may provide the comprehensive underlying theory that describes and explains our world, perhaps fairly soon. Although such a claim might seem controversial to many, I hope to convince the reader that after progress in recent years this is now a defendable goal, and one deserving of broad encouragement. I also hope to convince the reader that string theories about our world are testable.

1.1 The large number of solutions to string/M-Theory is not an obstacle to finding a Theory that describes nature

It is not necessary for the reader to know what string theory is or what M-theory is before reading the book.

This book is not a systematic review, nor a pedagogical explication. It is an enthusiastic, somewhat speculative, somewhat personal view of how compactified string/M-theories—plus some data that may be reachable—seem to have the possibilities of leading to a comprehensive underlying theory of particle physics and cosmology, perhaps soon. The book is well founded on three decades of compactification research in string theory, and over two decades of compactifying M-theory, which is used as the main example because it is where my own work has focused, and it seems to give very promising results. I will explain 'compactification' below. If string/M-theory is to be of value in understanding our world it will be through compactified string/M-theories—this book is about them. The study of compactified theories is largely called 'string phenomenology'. While the book is somewhat technical in places, I've tried to explain topics so that any curious reader can see the point of technical aspects even if not the details.

[In order to keep most parts of the book as self-contained as possible, I have not worked hard to edit out redundancy. Also, while some of the book is rather technical, the reader can obtain the sense of the arguments without trying to grasp the technical aspects. Some are in square brackets […] as a warning to avoid them unless you want a more advanced treatment.]

Many things fall into the string theory domain, and most of them are not directly relevant to explaining our world.

How would we know if we had a theory that described and explained our world?

We will see that we can make a list of about 20 questions (chapter 6) and issues, such that if we had a single theory that dealt with all or nearly all of them we would be reasonably confident that the traditional goals of physics had been achieved. Experts would largely agree on that list, although of course not everyone would precisely agree. We will see that there is a theory where most of the questions have indeed been addressed, and answers to basically all of the questions and issues in that example have already been achieved. The theory must be formulated in eleven space–time dimensions (11D). One dimension is time and the rest are space ones. We apparently live in four space–time dimensions (4D). This must be incorporated in our criteria for a legitimate theory.

We know our world has a gravitational force, one that is classically well described by Einstein's general relativity theory (GR), and that atoms and subatomic phenomena are well described by quantum theory.

Einstein's special relativity is well tested and established and we accept it, recognizing that space and time become intertwined. We won't need to know any details of how space–time works. Therefore the full 10D or 11D theory of our world *must* be somehow projected onto the four space–time dimensions.

In fact, the serious physics study of having more than the dimensions we explicitly see began about a century ago with the so-called 'Kaluza–Klein' ideas aiming at unifying electromagnetism and gravity by writing the theory in a 5D world (four space and one time). I will describe this more in the book.

All of our analysis in this book will be for the case where the extra dimensions are curled up into a small space–time region of approximately Planck-scale size, a region that mathematicians call a manifold under certain conditions. Terms such as space–time, Planck scale, manifolds, etc will be defined later in the text. Chapter 2 is devoted to the Planck scale. Such projected theories are called 'compactified string theories', compactification being the process that projects the ten or eleven dimensions onto four. Later we will focus on and provide examples for a particular one, compactified M-theory, since it has been well studied and has significant successes.

Any research area that is called string theory or M-theory but is not compactified to four space–time dimensions is not explicitly relevant to the subject of this book. As is often said, its successes or failures are not testable and have no direct connection to this book. We will see that some compactified string/M-theories generically behave like good candidates for a theory describing our world, and continue to do so as they are studied in increasing detail. There are a number of successes, some qualitative and some quantitative, and additional predictions, from compactified string/M-theories. I will describe some of them in this book.

What could be meant by testing a 10D theory without compactification to a 4D world? No one has given any useful meaning to that notion. If anyone claims

10D/11D string/M-theories are not testable, ask them what they mean, or what such a claim could possibly mean, since experiments are performed in a 4D world. Once one compactifies, the theory *is* testable, as we will see.

When Shakespeare wrote, there was no understanding of the physical world we find ourselves in, not one natural aspect of our world was explained. For reasons we probably understand, developments began about five centuries ago that led to our now having essentially a complete *descriptive* understanding of our world, of the world we see and of which we are all aware. Many things are explained. Amazingly, there is nothing about the natural world known from before that era that is now taught to students, but there are a number of things learned and discovered soon after that time which are still taught [1]. It is important to distinguish a descriptive understanding from an explanatory understanding. Some aspects of our world are now explained (for example what light is), and some are so-far only described (for example, parity violation).

From what has been understood, we have also deduced some things about our world that we do not explicitly 'see'. Some such things are surprising and/or counterintuitive. Probably the most obvious is that the Earth actually orbits the Sun, even though it 'obviously' does not.

As physics developed in past centuries we have accepted counterintuitive things about the world. The earliest one is probably our understanding of forces. Aristotle's writings have forces needed to make things move. As soon as the force stopped pushing, the motion stopped. Only with Galileo, and then Newton, was the correct idea formulated; an object will continue in its state of motion indefinitely until a force is applied. (The Aristotelian idea still applies with bureaucracies.) An early one, and perhaps the most amazing one, is that the Earth orbits the Sun, which is true but seems wrong every day. A surprising fact is that there is electromagnetic radiation we cannot see. Set up a prism so it shows sunlight as a visible spectrum of colors. Then put a thermometer where the visible spectrum ends on either side, and the temperature it shows rises—there is electromagnetic radiation we cannot directly see.

Perhaps another counterintuitive idea is true, and there are also additional space dimensions we cannot directly see, because they are too small. In the 1980s Michael Green and John Schwarz showed that to have a mathematically consistent quantum theory of gravity there must be nine (or ten) space dimensions (plus the usual one time dimension). That is a fact we should include in our thinking about progress toward an underlying theory. The natural size of the extra dimensions is the Planck size, so it makes sense that we cannot directly see them. (The difference between nine or ten space dimensions is a technical one we can ignore.)

It turns out that the extra dimensions can play a crucial role. The argument is a little technical, but one can see the general idea without understanding the details. We of course have three large space dimensions and one time dimension, so six (or seven) curled up ones.

Probably the most surprising and important of those are the very strong arguments that we live in a world with more than three space dimensions. As already mentioned above, these arguments have been exciting for nearly a century, embodied earlier in Kaluza–Klein theory and then in modern string physics which began in the mid-1980s.

After four centuries of study, our grasp of the world we *see* is almost complete. In order to say we understand our world, at least three things are necessary. First, we have to know what particles make up what we see, particles such as electrons and quarks (which are similar to electrons). Chapter 17 briefly describes how the families emerge from E_8.

Second, we have to know what forces bind the quarks into protons and neutrons, bind protons and neutrons into nuclei, and bind the electrons to nuclei to make atoms. Three forces (electromagnetic, strong, weak) plus gravity account for our world. If we knew the particles but not the forces we would not be able to describe or explain what we see.

Third, we have to know the rules to calculate the effects of the forces on the particles. Classically, for understanding motion, the 'rules' are Newton's second law, $F = ma$. The modern formulation of the rules combines Einstein's special relativity and quantum theory into 'relativistic quantum field theory'. Relativistic quantum field theory was written about 1930. It has not changed since then, and is not expected to change, although understanding of it has greatly increased since then and continues to improve. There are strong arguments that the rules will not change. Newton's second law works for any force, and so does quantum field theory.

One way this is sometimes described is to speak of an ultraviolet (UV) completion. In the electromagnetic spectrum the ultraviolet radiation is the most energetic, and the theories are most difficult at high energies, so we speak of needing the UV completion.

The combined knowledge of the particles, the forces, and the rules is called the 'Standard Model of particle physics'. It describes the world we see remarkably well. In its domain it is here to stay—it will not change. It was (essentially) completed by the discovery of the Higgs boson in 2012 at the LHC, at the European Laboratory CERN in Geneva, Switzerland. We will want to learn some aspects of the Standard Model as we go along in this book, but we will not need to know very much for our purposes. It is worth remarking that although it is called a 'model', the Standard Model is a full mathematical theory, the most complete ever written. The Higgs boson discovery is one of the most important scientific discoveries of the past century, because it points toward moving from a descriptive theory to an explanatory underlying theory that includes particle physics and cosmology.

There is also a Standard Model of cosmology. It includes Einstein's general relativity, and gives a good detailed description of a universe that first was of Planck-scale size with an unstable energy density that underwent an initial very rapid growth in size, an 'inflation'. After a short time the unstable energy density transitioned (the Big Bang) into a large number of energetic particles and some remaining energy density. The Universe has been expanding since then.

The Standard Models describe our Universe in terms of a relatively small number of parameters. It is very interesting to think about what we mean by 'small' here. In one sense the Standard Models are said to have about 30 parameters, such as the electron mass or the strength of the gravitational force, and that seems like a lot.

Ultimately we hope to have a final theory with no adjustable parameters, or very few. Later in the book we will see how that can arise. Actually 30 is probably an exaggeration, because when we understand quark masses we will probably find that the theory of quark masses explains several of them simultaneously, so those should really have been thought of as one. Similarly, there are four force strengths in the 30, but if the forces are unified then probably those four are all determined at once— they just look different.

On the other hand, historically physics was studied as a number of separate topics, such as motion, sound, waves, heat, thermodynamics, electricity, magnetism, and many more. Each of them was formulated so that it had a few input quantities, from which the rest of the phenomena in that area were calculated. The total number of parameters originally needed to describe all the phenomena encompassed by the Standard Models would give many tens, maybe hundreds—all the currently needed ones plus lots more that are now calculable in the Standard Models. So actually there has been great progress from the point of view of consolidating parameters. Even so, we expect (or hope) to eventually obtain a final theory of our world with zero or very few adjustable parameters, which will be described in this book.

In my view, we are living during a hugely exciting era for physics and for science, and for people more generally, one during which it may be possible to finally achieve a real understanding of our physical world, and the sense of dignity and meaning that could come along with that understanding. Two things arc crucial for that to be so, one experimental and one theoretical—either alone would be less exciting.

First, experimentally, the facility at the European Laboratory CERN called the LHC that collides beams of protons to create new particles is finally able to take data at energies and event rates where well-motivated theories suggest new particles may be observed that point to how to formulate the underlying theory. Some people might say such claims have been made or could have been made in the past so why take them seriously now. The difference is that now the claims are based on calculations in actual theories, while in the past they were based on analogies or 'naturalness' arguments. More broadly, the Standard Models do not leave descriptive gaps or puzzles. Additionally, it was learned in the 1980s that dark matter should be composed of new forms of matter, not of the atoms from which we are made. Finally, after over three decades of development detectors are beginning to operate that are sensitive to most hypothetical forms of dark matter in realistic amounts, and new detectors are being designed to cover any regions that cannot yet be searched. Earlier dark matter detectors might have detected the dark matter (or part of it). Searching for dark matter is now a mature area.

Without the LHC, a facility supported regionally but not by any single country, and the dark matter detectors developed recently, and perhaps the next generation of dark matter detectors, we might never have the data to point to and confirm an underlying theory. Of course it may be that the discoveries needed will not be made, but at least we know we are in the region where optimism is reasonable and defendable. **No amount of cosmology or astrophysics theory could tell us how to complete or to extend the Standard Model, or what the dark matter is. Experimental discovery is crucial.**

Actually, the situation today is surprising to many because almost all the data are in, so to speak. We know: the Universe is long-lived; there are three large space dimensions; the amount of dark matter; the size of the matter asymmetry of the Universe; the Universe is geometrically flat to high accuracy; the dark energy equation of state is essentially unity; the size of the cosmological constant; that the rules of quantum theory hold; that an effective theory of inflation gave an accurate description of data including the high scale of inflation; electron and neutron electric dipole moments are surprisingly small; there are three families. Quark and lepton masses, and soon even neutrino masses, and the Higgs boson mass and decay branching ratios are approximately measured and soon will be accurately measured. Even though several of these items are technical, the reader can see the point that most of what we need to know is known. The two big gaps in data are what particle(s) make up the dark matter, and whether there is electroweak-scale supersymmetry (defined precisely later).

For historical reasons CERN has a long-term treaty-based budget from member countries, rather than annual fluctuating funding from countries with dysfunctional congresses, so eventually CERN will have the funding to build facilities with new levels of energy. China has begun to discuss building a more energetic collider without requiring help, which would be a wise decision for several reasons for a leading country. While the costs are large, the investment would be paid back many times over in economic and human returns. An international R&D program is underway to plan a circular proton–proton collider whose energy is up to seven times that of the LHC. Calculations imply that such a collider would make important discoveries. It should provide the data to definitively solve the Hierarchy problem (see below). There are now conceptual design reports that would be a basis for proceeding.

Dark matter detectors, while not inexpensive, are being built by several countries. Again the investment would be well rewarded in economic terms, even without discoveries. Dark matter detectors are rather specialized so several kinds may be needed. As we will see later, for dark matter there is currently little theoretical guidance. Altogether, on the experimental side we have good scientific arguments that the discoveries needed to confirm or disprove the theories should be possible in coming years. However, if we had all the data but no theory, we would not and could not know what the data implied.

Think about a theory that really explained the Universe, its origin and all the variety in it, from stars to atoms with life in between. We would need to write some equations to calculate things. We would need expressions for quantities having dimensions. We would want the quantities in the equations to only depend on rather universal measures, such as Planck's constant h for the size of quanta, Newton's constant G for the strength of forces, and Einstein's constant c for the speed of light. There should be no dependence on quantities whose dimensions are artefacts of our world, such as electron volts or a foot.

From h, G, c we can construct a quantity with the dimensions of length, called the Planck length, which could be the typical diameter of a universe. This turns out to be $(Gh/c^3)^{1/2}$, which surprisingly is about 10^{-35} meters. Nevertheless, that must be the natural size of a universe. Universes which differ much in size from this (e.g. ours) need to be explained. The unit of time is about 10^{-44} sec, so universes that live much longer than this, such as ours, need to be explained. The Planck energy scale is about 10^{19} GeV, or about 10^{-8} kg. In our world the mass scales are set by the Higgs boson mass, about 10^2 GeV, or about 17 orders of magnitude that needs to be explained. All dimensionful quantities can be expressed in terms of h, G, and c. We expect the short-lived curled up regions of space arising from compactified theories to be about the Planck size in diameter. The details of units don't matter.

The string theory framework has all the richness and structure needed to provide an underlying theory. It is easy to misunderstand what that statement means—a significant part of this book is devoted to explaining it. It is remarkable that Michael Green and John Schwarz figured out in the mid-1980s that to have a mathematically consistent quantum theory of general relativity describing a world meant having a world with nine space dimensions. When the extra dimensions are curled up into a Planck-scale size manifold the resulting compactified theory generically behaves like one with strong, electromagnetic, and weak forces in addition to the gravitational force described by general relativity. The resulting theory behaves like the theory that has increasingly been formulated piece by piece over the past century. It is a coherent, consistent theoretical framework that addresses all the basic questions physicists and cosmologists have wanted to ask about our world. Much has been written about the testability of string theories—we will see that *compactified* **string/ M-theories and only compactified string/M-theories are indeed testable in the traditional way of physics theories**, contrary to what is being said and written in a number of journalistic articles, blogs, and books.

String/M-theories have many solutions, a landscape of them. An early estimate was of order 10^{500}. It was frequently said that with so many solutions how could one hope to find our world! This kind of argument is misleading in two ways. First, simple requirements exclude almost all the solutions as candidates for a world. For example, none of the 10^{500} models contain an electron (M Douglas, private communication), which is needed for an acceptable universe, so they could not produce viable universes. The relevant aspect is the large number.

Only solutions that undergo inflation, which is unlikely, can live long; all the rest are excluded. We'll see below that **finding solutions like our world is not hard.** The situation is somewhat analogous to quantum field theory. There are an infinite number of consistent relativistic quantum field theories. A few experiments allowed picking out the Standard Model as the correct one. It is similar for the string/M-theory landscape. Some measurements are needed. Then many predictions can be made. We will see that M-theory provides an example. **The large number of solutions to string/M-theory is not an obstacle to finding the correct theory. The compactified theories look like theories of our world.**

There is some research on purely gravity-based theories, so-called loop quantum gravity, or emergent gravity. Because such work intrinsically does not have any connections to the other forces, or to the particles, we will not consider it further in this book. That the string/M-theories address all these aspects (all the forces, the particles, and the rules) simultaneously is one of their important strengths.

Several exciting features arose as the resulting theoretical framework from compactified string/M-theories was studied over the past three decades. **The Standard Model is a special kind of quantum field theory, called a Yang–Mills gauge theory. It turns out that compactified string/M-theories generically are such theories. In addition, it turns out that compactified string/M-theories also contain states with properties like those of massless quarks and leptons, again a property of the Standard Model** (before the Higgs physics plays its role and allows small masses compared to the Planck scales). These are powerful tests of string/M theories. The compactified string/M-theories could only describe the real world if they had these properties, and they do, so they have passed powerful tests already. **These generic features make it much easier to find them.** Ultimately a particular Yang–Mills gauge theory with the Standard Model particles and forces must emerge if this approach is valid. The situation today is encouraging, although of course not guaranteed to succeed.

Supersymmetry is a possible hypothetical property of an underlying theory, a very desirable one. It has a number of attractive features, one being it allows connecting the Planck-scale ideas with the Standard Model. Whether nature is indeed supersymmetric should be testable at the LHC, and is probably the most exciting goal of LHC data in the coming years. It would be discovered via observing some new particles related to Standard Model particles (superpartners). People had hoped that such discoveries would already be made, based on naïve reasoning (called 'naturalness') rather than serious calculations, but superpartners have not been found so far. That has led to superficial claims that nature is not supersymmetric. Actually, **LHC is just entering the well-motivated region of energy and intensity where compactified theories imply the superpartners should exist**. That was not known before a few years ago, because we had not learned to calculate masses in compactified string/M-theories before that.

It's worth spending a little more time thinking about the absence of apparent superpartners. The key is actually a mystery called the 'Hierarchy' problem. It's the central problem of particle physics today. It's a subtle problem. There is no experimental contradiction. The Standard Model explains and describes particle physics at the 'electroweak scale', the scale of W's and Z's and tops, and all their interactions.

Basically the Hierarchy problem is simple. **We expect the natural sizes and lifetimes of universes to be the Planck scales, 10^{-35} cm, 10^{-44} sec, and 10^{18} GeV (energy or mass units). So the very different values of our Universe need to be explained. As we'll see, inflation can explain the size and lifetime. All the quarks and leptons and 'gauge bosons' (that transmit the forces) are massless in the SM theory. They get mass from the 'Higgs mechanism'** (we'll describe the Higgs mechanism later). The quantum of the Higgs field, the Higgs boson, gets mass also. The masses of the other particles are proportional to the Higgs boson mass. **All this works well, and leads to a satisfactory theory.** The masses of the Higgs boson, the top quark, and the W and Z bosons that mediate forces are all about the same, of order 10^{-17} of the Planck mass; the remaining masses are even smaller. **So we have a description of the world with activity at two scales, the Planck scale and the much smaller so-called 'electroweak scale'. Our world (atoms, molecules, etc) is at the electroweak scale. That would be all right if we could deduce the electroweak scale from the Planck scale.**

That's where the Hierarchy problem emerges. In a world described by quantum theory, calculations of observables have a leading term, plus quantum corrections. The quantum corrections depend on virtual particles. It turns out that the quantum theory corrections to the calculation of the Higgs boson mass depend on some of the Planck-scale particles, and bring the two scales together. In the Standard Model it is not stable to have the two worlds separate!

In 1992 Steven Weinberg wrote 'for over fifteen years the Hierarchy problem has been the worst bone in the throat of theoretical physics'. 1992!

Supersymmetric theories were discovered by theorists in the 1970s as elegant and interesting theories. They were not invented to solve any problems—in fact, for a few years after they were discovered they were often called 'a solution looking for a problem'. Around 1980 it was recognized that in a supersymmetric world the Hierarchy problem would be solved! That was because the troubling quantum corrections were canceled by super-quantum corrections. In order to claim the Hierarchy problem was indeed solved, it is necessary to confirm that the super-partners predicted by supersymmetric theories indeed existed. The LHC has so far just entered the region of superpartner masses predicted by compactified theories, which ranges from about 1.5 to $\sim$5 GeV (we'll discuss that range later). Those values are the only physics predictions, rather than just speculations. The LHC will run with higher luminosity after an upgrade, beginning in late 2021 if pandemic work stoppages do not delay it. That increases the possibility of discovery, though not very much. A higher energy collider is needed. From what we know now, a collider with twice the LHC energy range would probably suffice, and cover the region of gluino masses to about 5 GeV.

The issue is extremely important. **If we are ever to realize the goal of understanding our Universe it will depend on formulating an underlying theory in the natural Planck-scale region. Then it will have to contain a solution of the Hierarchy problem and allow us to calculate our atomic part of the world, physics at the electroweak scale.**

Even if we are lucky and superpartners (or some other new physics that allows a solution of the Hierarchy problem) are discovered at the luminosity upgraded LHC, **we will need a higher energy collider to test the ideas about connecting the Planck scale and our scale. To not build a higher energy collider (two or more times LHC) would be giving up on the age-old quest to fully understand our world.**

Fortunately, the CERN budget is based on treaty agreements because it came into existence soon after a war which left the member countries not trusting each other. It is also indexed to inflation. **Eventually CERN will build a higher energy collider. China may also take this opportunity to become a world leader in particle physics.**

String/M-theories have many solutions, which has led to the notion of a 'land-scape' of universes. Some people have been concerned about how we could find a solution that described our world if there were very many to examine. Since most string theorists do not focus on compactified theories, which we have seen already are very much like our world, they are confused about this question, but it is not a problem, as we will see. **Since the compactified theories generically are like the Standard Models, it is not hard to find good candidates.**

Another subtlety about the landscape is that many of the solutions will not be populated universes—e.g. they will only live very short Planck-scale times. This has been studied a little, but its implications are not yet understood. **It's easy to recognize interesting universes—if they inflate they are good candidates, and vice versa.**

1.2 String theorists study theories, not phenomena

Much has been written or said in praise of, or criticism of, 'string theory'. Most criticisms apply to parts of string theory that are not relevant to explaining our world. **For centuries the goal of physics was understanding the world we find ourselves in. Now that goal may be in sight via the combination of string/M-theory plus data from the Large Hadron Collider (LHC) and on dark matter. But sadly, research aimed at that goal has largely been abandoned by string theorists.** That is clear if one looks at the talks they have at conferences, at seminars at their universities, and in PhD theses. The program for the annual international conference on String theory illustrates well how none of the talks try to address real world issues: https://indico.cern.ch/event/929434/timetable/.

Why don't string theorists make theories of the real world? One can only speculate, of course. If string theorists are asked whether the discovery of super-partners would affect what they work on, the typical answer is no—they already assumed superpartners were there, and confirming it was nice, but they will keep working on what they were working on. Working on the real world was complicated, and working on the more well defined and mathematical problems of pure String theory was nice. They had momentum there. That's more or less what happened with Higgs physics after the discovery of Higgs bosons.

1.3 Nutcracker

Today, both the experimental and the theoretical parts of our quest for extending and deepening our comprehension of our Universe are ripe for fruitful progress. Either without the other is unlikely to take us to a new stage; with both good progress is possible, perhaps likely. Figure 1.1 is an image that I hope will easily come to people's minds. The object about to be cracked is a projection based on a Calabi–Yau space (provided by Andrew J Hansen of University of Indiana). **If only the bottom arm of the nutcracker (data and phenomenological theory) is used it will not crack open the 6D object so we can get at the underlying physics and theory.** Similarly, if only the top arm is used without the bottom (string theory) it will not be possible to get at the **underlying physics and theory. But if both are used it is at least possible, and perhaps now probable**.

Finally, I want to repeat and emphasize that compactified string/M-theories are strong and testable candidates for theories that provide a comprehensive underlying theory that describes and explains our world, incorporating the Standard Models of particle physics and cosmology. The purpose of this book is to explain and document that statement. As we will see, the compactified string/M-theory perspective changes how we view the world in a number of ways, even if we only study Standard Model physics and related issues for understanding our world.

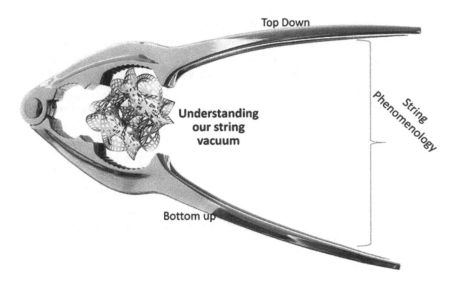

Figure 1.1. The nutcracker approach to understanding our world. This figure illustrates that learning the physics and predictions and tests in the curled up dimensions requires both a top-down (theory) approach, and a bottom-up (data, phenomenology) approach. With either one we cannot expect to crack open the (here projected Calabi–Yau) manifold shown as the nut (drawn by Andrew Hansen), but with the combined approaches we can. A similar diagram with a 7D G_2 manifold curled up could eventually be drawn for compactifying M-theory.

1.4 Why the Standard Model is true—it will be extended but not replaced in its domain

We know the Standard Model of Particle Physics describes our world. It is an awesome theory. It describes what is in us, and the world we see, completely. It tells us what the fundamental particles are (electrons and quarks—quarks are like electrons but carry an extra charge different from electric charge). It tells us the fundamental forces are the electromagnetic force, the weak force, and the strong force, and how to calculate the way the forces act on the particles. The strong force binds quarks into neutrons and protons, and a residual strong force binds protons and neutrons into the nuclei of the chemical elements. The electromagnetic force binds electrons to nuclei to make atoms. Although it is called a 'model' for historical reasons, it is a full relativistic quantum field theory, the most mathematical and complete theory in history. When joined with classical gravity it accommodates and describes all that we see.

We are confident the Standard Model is here to stay. It is true, and will not change, contrary to what lots of philosophers and historians say. It will be extended but not replaced. Classical electromagnetism is described by Maxwell's equations. **The Standard Model contains Maxwell's equations—they are extended to the Standard Model, which joins them with the weak interactions to give the electroweak part of the Standard Model.** A textbook on the Standard Model written recently will not need a new edition. Sometimes we say we 'believe' it is true and here to stay. It is important to understand that 'believe' is used in a special way here, to mean there are strong theoretical and experimental reasons why we are confident the Standard Model is here to stay, and a number of expert physicists I know and trust have examined all issues where something could go wrong. Belief is not used here in the sense of having faith.

The Standard Model is not a 'theory of everything', in two senses. The Universe has aspects we don't see yet, such as dark matter that is deduced indirectly from its gravitational effects, or perhaps a supersymmetric extension. Second, it's important to distinguish between the theory and its solutions. For example, **the fundamental equations of condensed matter physics are part of the Standard Model, but solving them for complex systems is so difficult that emergent behavior (such as life) arises but could not have been predicted**. The Standard Model may be extended to be a part of a more comprehensive theory, such as a supersymmetric Standard Model or a compactified string/M-theory, but it will not be changed in its domain—and its domain is everything we can see.

There are two kinds of reasons why we believe (in the above sense) the Standard Model is here to stay. The first reason is that it leads to many predictions that have been confirmed by experiment. I'll describe some below. And none of its predictions in its domain have been falsified.

A number of its predictions are very strong and unusual ones. Confirming experimental predictions is not a fully satisfactory kind of verification, because an improved test at a new facility that probes the theory more deeply could fail to confirm it tomorrow. There are two theoretical arguments that imply the Standard Model is true, arguments that are very general and not subject to change tomorrow. I'll describe them below.

1.5 Experimental confirmations of the Standard Model

There are many (hundreds) of quantitative tests of Standard Model predictions. Rather than enumerate a list of them, I will describe two powerful tests here.

- First, many Standard model predictions depend on a particular parameter. Call it X. For example, the probability for scattering a muon neutrino and an electron is proportional to a factor $((1/2 - X)^2 + X^2/3)$ while the probability for a Z boson to decay into an up quark and an anti-up quark is proportional to $(1/4 - 2X'/3 + 8X'^2/9)$. In the Standard Model $X = X'$, while without the Standard Model X and X' are unrelated. Further, in the Standard Model X must turn out to be a real number between 0 and 1, while without the Standard Model X and/or X' could be any different numbers, positive or negative or even complex. And there are about ten such factors, with X and X' and X'' etc. In fact, all the X's turn out to be equal and a positive number between 0 and 1, $X = 0.23$. This is a very powerful test.

- Second, the Standard Model predicts the existence and some masses of a number of new particles! It predicts the existence of eight gluons that mediate the strong force, and the W boson and Z boson that mediate the weak forces and their approximate masses, and the top quark. All were found to exist with the predicted masses and spins and electric charges. This is a different kind of test, and again a very powerful one.

Neither of these tests is of the sort that could break down as more data comes, so they strongly encourage us to be confident the Standard Model is here to stay. Nevertheless, epistemologically such empirical confirmations have an intrinsic weakness with respect to future data.

1.6 Theoretical evidence for the Standard Model

- First, the Standard Model satisfies all the requirements to be a full theory. It is valid to arbitrarily high energies, and at very low energies. It satisfies the conditions of Einstein's special relativity and of relativistic quantum theory. No previous attempts to formulate theories were able to achieve such goals. There can be no theoretical limit or constraint in which the Standard Model breaks down, except for the Hierarchy problem. It has no puzzles or hints of a problem. This is the first time in history there has been such a theory. It is a strong candidate for a comprehensive theory.
- The second theoretical argument depends on the property of any relativistic quantum theory that the strength of forces depends on the energy at which they are probed. At our energies, we know that the forces have different strengths, so the goal of unifying them doesn't seem to make sense. Newton speculated that there must be more forces, and that they would somehow unify. Two hundred years ago Coulomb showed that the gravitational and electromagnetic forces have the same form, though very different strengths. From then on, many physicists have aimed to unify the forces in form and strength. Einstein spent half his career focusing on that, without success because he didn't know about the weak force or quarks. The Standard Model can achieve that goal qualitatively.

Let me explain a little more about how the Standard Model encourages us to see the forces as unifying even though at our energy scale the forces are of quite different strengths. As the weak and electromagnetic forces are probed at higher energies their strengths become more alike, and those two forces unify into the 'electroweak' force at a scale not far above the W and Z boson masses. Probing at even higher energies the strong force decreases toward the electroweak force. In the Standard Model the forces never quite become the same, though if the Standard Model is extended to the supersymmetric Standard Model then the strong, electromagnetic, and weak forces do have the same strength at a high scale, only a little below the Planck scale, the natural scale of gravity, so unification with gravity is then plausible.

We know the Standard Model is here to stay, because of combined experimental and theoretical reasons. But that is not the whole theory.

Although we know the Standard Model is a true mathematical image of the visible world, we also know it is not the whole theory. We know the latter because the Standard Model is only a descriptive theory, it does not *explain why* the electron and quarks are the fundamental matter particles, or why its forces are the actual ones. Second, it does not explain some things we do not directly see, such as the presence of dark matter. We expect an underlying comprehensive and explanatory combined theory of particle physics and cosmology, in which our world is described by the Standard Model. Such theories can arise as string/M-theories with nine or ten space dimensions are compactified to three space dimensions, and are being actively studied. We will describe them below.

1.7 In recent decades the boundaries of the goals of physics have changed

Starting in the 1970s with the development and confirmation of the Standard Model of particle physics, and the discovery of the idea of supersymmetry, and approaches about unification of the electromagnetic, strong, and weak forces of nature (called Grand Unified Theories), and then greatly encouraged by the discovery of inflation and of string theory in the 1980s, previous conceptual limits about what people thought could be understood about nature began to be ignored. For the first time in modern history, physicist's thinking and publishable research began to include the possibility of a unified, underlying comprehensive theory of the physical world, a 'final theory' based on publishable, testable physics research, not on philosophical speculation.

That is not to say that there is already such a theory in place, but that increasingly many physicists began to take seriously the possibility that the ingredients needed to achieve such a theory might be already or soon available rather than at best a goal for the distant future. I will argue here with concrete and detailed examples that is indeed the case. I will even argue that we may be much closer to achieving a final theory than is recognized by many workers, even those close to the subject.

Before some readers get excited and disagree, we need to define 'final theory'. We mean *only* a final theory, not a theory of everything. We use 'final theory' in the sense of Steven Weinberg [2]. The final theory will replace descriptions with explanations and deeper understandings, and propose answers to questions such as what is a quark, why are there three families and the symmetries of the Standard Model, what is the physical inflaton that generates inflation, and what is the cause of the matter asymmetry. All these answers and explanations must emerge from one comprehensive underlying theory, a theory we can call the final theory. This optimism arises from the situation with current and past research and ideas, and with the newest experimental facilities.

The quest may not succeed. The optimism is useful because it leads people to work harder on the relevant research. The final theory does not even try to explain the black hole information paradox, or higher spin theories, or AdS/CFT (the reader does not need to know what these are), etc. If no final theory (or almost final theory) emerges, eventually people will concede that and largely lose interest.

Many experts who specialize in various aspects of string theory will not endorse the possibility that the final theory of our vacuum may be soon forthcoming, because they work in technical areas that lack overviews. Anyone who focuses on solutions in other than four dimensions, or black hole solutions, or anti-de Sitter space/CFT, or amplitudes, or higher spin theories, or dualities, or holography, or moonshine, or a few of these, or many other areas will have no reason to expect a comprehensive theory to emerge. And of course solutions can be constructed that do not describe our vacuum. Similarly, experts in QCD physics or Large Hadron Collider (LHC) physics or dark matter physics will also not have an overview of the ingredients described below, and generally will not be aware of the emerging final theory opportunity. Those who are not aware of it will typically be skeptical about it.

In this book we focus on how we can recognize if the quest might be successful. How will we test it? First we discuss the issue of testing theories in physics in general, and clarify some aspects. Then we describe several ingredients that seem necessary for progress, such as the idea of supersymmetry. Next we give a list of questions that should be answered by a final theory. This list is interesting, and people like looking it over and checking it against their ideas and goals. We will illustrate reasons for optimism by giving an example where all the questions are already answered in one theory, compactified M-theory, and describe a number of its tests.

Those who do work toward a final theory typically have done so via a 'top-down' approach. They imagine writing a ten-dimensional (10D) string theory, finding that some principle leads them to a particular compactification to four dimensions, and a particular vacuum state. It would be fine if that worked, but so far it has not been a fruitful way to proceed. That is not surprising since historically physics has always been driven by both theory and data. In addition, if any crucial aspect of data, or any concept, is missing, progress can be greatly slowed. What works well is what I have called the nutcracker approach (see figure 1.1 and its caption). If the end result is viewed as cracking open the underlying curled up small dimensions structure, then both the top-down and the bottom-up (data driven phenomenology) approaches will be needed.

The Standard Models

As described in the preface, if we want to achieve an understanding of the physical Universe, at least three things are necessary: we must know the basic constituents everything is made of, the particles; we must know the forces that bind the constituents to form our world; and we must know the rules to calculate the results of the forces acting on the constituents. Historically the rules came first. The rules are Einstein's special relativity plus the quantum theory, combined by 1930 into relativistic quantum field theory. Quantum field theory is a framework that holds for any forces and particles. During the second half of the twentieth century the particles and forces were identified.

The particles that form all we see are the familiar electron, and two more called quarks (named the up quark and the down quark, for a reason). Quarks are like electrons, but carry an additional charge (called the 'color charge') so they feel an additional force (the 'strong force' or 'color force'). The use of the word 'color' is an analogy; there is no connection to real colors. The up and down quarks bind to make protons and neutrons, which have no net color charge. Protons and neutrons bind into the nuclei of the chemical elements via a residual leakage of the color force, called the nuclear force. Electrons bind to nuclei via the electromagnetic force, making atoms. Atoms are neutral but bind via a residual leakage of the electromagnetic force into molecules. Molecules form flowers and espresso and people and planets and stars, and all we see.

This picture is fully quantitative and the results are calculable at all levels, although sometimes solving the equations gets to be so complicated it has not yet been done. It is the simplest description ever to emerge, simpler than the attempts of the Greeks or other cultures, such as earth, air, fire, and water. And it can describe or explain all that we see.

The quanta of the electromagnetic force that bind electrically charged particles into atoms and molecules are photons, and the quanta of the color force that bind colored particles into hadrons are gluons. Essentially all the quarks and electrons in everything we see were created in the Big Bang. There are several other kinds of quarks, and other particles like electrons, and more particles, but they do not enter directly into what we see.

Four forces are necessary and sufficient to form the particles into the world we see. The Standard Model incorporates the electromagnetic, strong, and weak forces in a fully relativistic quantum field theory. Classical gravity is attached.

Particles have another property, called spin. In quantum theory spins are quantized, so they can have spin 0, 1/2, 1, 3/2, 2, all in units of Planck's constant. Physical theories with larger spins do not occur. In quantum theory spin is analogous to angular momentum, but not precisely. In practice we will not need to do more with spin than be aware of it being a property of particles. Particles with integer spins are called bosons, and particles with half-integer spins are called fermions. For individual particles their behavior is not particularly different if they are bosons or fermions, but two or more that form a system do behave differently, sometimes dramatically so. That will not affect us except occasionally indirectly, for example as to whether a new kind of particles called axions form dark matter. Supersymmetry is a symmetry of the theory under which bosons and fermions in the theory are interchanged. Matter that we see is made of fermions, while the bosons mediate the forces.

The theory called the Standard Model of particle physics is an awesome theory. The name 'Standard Model' is of historical origin, and is misleading in that the Standard Model is the most mathematical theory known. It describes the world we see, and explains much of what we see. It achieves the goals of physics since its modern beginning four centuries ago. The Standard Model is very well tested, and in its domain it is here to stay. To honor it we capitalize its name. A textbook written now on the Standard Model will never have to be updated.

We add the Standard Model of cosmology to that of particle physics. We know rather well what comprises the Universe: about 5% is the matter particles we see and are composed of, mostly in the form of atoms. About 25% is dark matter, some sort of particles that clump under gravity as normal matter does, but do not form stars and emit light. We do not yet know what the dark matter particles are, although we know a lot about what dark matter is not. There are some good candidates for what constitutes the dark matter.

About 70% of the Universe is called 'dark energy' and does not behave as matter under gravity but has a uniform density throughout the Universe. This part is effectively a force that causes the expansion of space–time itself at an increasing rate. A small amount is radiation (photons)—about a part in 100 000. Some is neutrinos, about 0.5% or a little more.

There is also a standard cosmological history. There is good evidence that at an early stage the Universe was an unstable energy density, which inflated from a Planck-scale size to a size perhaps of the order of a soccer ball very rapidly, an increase in size of a factor of about 10^{35}! Then it decayed into energetic particles, in what we call the Big Bang. The history usually assumed (called a thermal history) is that after the Big Bang the energy density of the Universe was dominated by photons since most particles decayed or annihilated into photons fairly quickly ('radiation dominated'). It simply expanded and cooled. We will see that compactified M-theories imply instead a non-thermal cosmological history—this is one of their generic predictions.

The Standard Models together describe an amazing amount, but are actually incomplete at a conceptual level. It is helpful to state three ways they are incomplete, although the three are somewhat related. They are:

1. Some questions or phenomena have no Standard Model answers or are not described by them. For example: why are there more families of particles like up quark, down quark, electron (and electron neutrino), and why are there apparently precisely three such families? The additional families seem to play no role in the behavior of our Universe; their role must somehow arise in the underlying theory. And the Universe has a matter asymmetry, with over a billion times more matter than antimatter. At the Big Bang one would have expected the initial state of pure energy to turn into equal amounts of matter and antimatter. How does the matter asymmetry originate? And what is the physical inflaton that increases the size of the Universe such a huge factor during inflation? What is dark matter? Why does the strong force (quantum chromodynamics (QCD)) not have interactions that violate the conservation of charge parity (CP, the product of charge conjugation and parity trans-formations) even though such interactions are fully allowed in principle? The Standard Model cannot answer these questions and others; it is not that we have not yet figured out the answers in the Standard Model, it cannot.

2. The Standard Model is a descriptive theory, because it does not explain why its particular electroweak and color forces are what they are, and whether they are inevitable, or why the constituents are quarks and electrons instead of some other particles, or why the CP symmetry is slightly broken rather than conserved. One can extend the list of such 'why' questions.

3. The Standard Models are also **effective theories**. Fortunately, much of the physical world can be divided into isolated domains that can be studied separately, each called an **effective theory**. That is why physics is the easiest science. The physics of atoms assumes relativistic quantum field theory, the electromagnetic force, the existence of electrons and nuclei with specific masses and charges and spins, and then deduces everything in its domain.

For understanding atoms we do not need to know about quarks, or stars, or anything else. The Standard Model inputs quarks and leptons and their masses, the Higgs physics, the four forces (attaching gravity) and their strengths, and the rules of quantum field theory, and then deduces the existence and properties of protons, nuclei, atoms, stars, and so on. It does not need to know about inflation. **If a theory has inputs it is an effective theory. Eventually we hope for a theory without inputs, which can be imagined**.

If all the ingredients are in place, progress can be quite rapid. We have a good and relevant example of that with the Standard Model of particle physics. At the beginning of the 1970s people spoke disparagingly of the status of particle physics. But with the improved understanding of non-abelian gauge theories, plus the already existing data and theoretical structure, the whole picture fell into place and by 1973 the full Standard Model existed. It is arguable today that we are in a similar situation, with the missing ingredients being both experimental (confirmation of supersymmetry and learning what the dark matter is) and moduli stabilization plus supersymmetry breaking in compactified theories (explained later). The LHC has now started working in a region of energy and intensity where well-motivated theories imply superpartners could have been seen, but were not yet seen. Given the ingredients listed here, one can defend the point of view that with the discovery of superpartners and identification of dark matter a final theory of our vacuum could emerge quickly, with the data perhaps pointing to the appropriate detailed theory. LHC does not yet cover the full region needed for at least one superpartner, and we don't know the parameter region needed for hidden dark matter. Once the theory tells us to look at (say) decaying dark matter, the hypothetical model allow us to calculate the sizes of any possible signal and background.

Einstein spent most of the last two-thirds of his research career searching unsuccessfully for a unified description of the gravitational and electromagnetic forces. In hindsight we understand that his search could not have been successful, because he did not try to include unifying with the weak and strong forces as well, and because he did not know about quarks. A comprehensive unified theory would have to include several essential parts in order to work. It would have to be a relativistic quantum field theory in four space–time dimensions. It would have to include the particles and forces of the Standard Model of particle physics, and General Relativity.

Another necessary ingredient came well over a century ago. With the invention of quantum theory, Max Planck realized that physicists had discovered the three fundamental constants of nature needed to make universal units, the so-called Planck scales (explained below). A final theory would need to be formulated at length and energy and time scales that were universal and independent of people or accidental features. With the proportionality constant for force in Newton's law of gravitation, and the speed of light that emerged from Maxwell's equations, and Planck's quantum scale, it became possible to express any quantity having units in terms of these three quantities. One could imagine a theory having a simple set of equations expressed in Planck units. This is explained in more detail in the next chapter.

Supersymmetry (explained in more detail below) is another crucial ingredient, because it allows formulating a theory at the Planck scales, while the same theory implies other phenomena at the scales of our vacuum, where protons and nuclei and atoms can exist and form our world. With supersymmetry different scales can exist at stable separations because the supersymmetric contribution to divergent observables cancels the Standard Model one.

The idea of supersymmetry is well formulated, but not yet explicitly confirmed experimentally. There is good reason, based on theory, to think discovery of the superpartners of Standard Model particles should occur at the CERN LHC in the next few years. Without technological societies and institutions that pursue fundamental research goals and fund them at the very high levels needed, the supersymmetry ingredient might never be tested. Once supersymmetry was formulated its experimental absence would not have blocked formulating a final theory, but without the actual discoveries important tests would be lacking, and fewer people would be convinced the theory indeed explained our world.

In this book we will assume the final theory is a supersymmetric one, and that it provides the solution of the Hierarchy problem. There is good motivation to do that. There are several phenomenological arguments which will be described in more detail below. They include accurate unification of the Standard Model coupling strengths for SU(3), SU(2), and U(1); radiative electroweak symmetry breaking; the stability of the vacuum; the properties of the Higgs boson; and more. Experimentally, searches for superpartners will be among the first ones when data in an increased range of energies are available.

The final crucial ingredient was string theory. What is string theory? What is any theory? The main goal of physics has been to understand the physical world. That means writing a consistent mathematical theory that describes the physical world, understanding why that theory actually describes nature, and testing enough predictions and explanations to be confident the theory works. Historically one writes effective theories that increasingly encompass larger areas, integrating ones that cover smaller domains. Each effective theory has some inputs and derives others. Eventually we hope for a comprehensive one that has no, or almost no, inputs.

Given our world, it is essential that the final theory include a relativistic quantum theory of gravity. Amazingly, in 1985 Michael Green and John Schwarz showed that mathematical consistency required that such a theory, one that could also describe quarks and electrons, should have nine space dimensions. In that theory the basic objects would not be point-like as in quantum field theory, but extended. They could be stringy, e.g. extended in one dimension, or perhaps more complicated. Stringy objects are the simplest case, and much of the mathematics of stringy objects had already been studied, so that quickly became the default.

A decade later **Edward Witten showed that a related approach, called M-theory**, implied ten space dimensions and could also allow formulating a consistent quantum theory of gravity.

String theories provide a fruitful opportunity for much study. Obviously, in order to provide a framework for an underlying theory for our world, string/M-theories must be projected onto four space–time dimensions, a process called 'compactification'.

The projection is naturally achieved by making the other dimensions Planck-scale size. When that is carried out, one might worry that information is lost. On the contrary, remarkably **it turns out that the resulting theory has not only gravity, but also the electromagnetic, weak, and strong interactions. In addition the basic string massless states can be interpreted as quarks and electrons, and as the particles that mediate the forces, photons and gluons, and the W and Z bosons of the weak force.** Relativistic quantum field theory fails to provide a quantum theory of gravity for point-like particles. Treating particles as points is too singular. Probably any way of giving them extension would work; strings are just the simplest case.

Sometimes one reads descriptions of string theory and particles as vibrating strings. While correct, that is actually not very helpful. The particles we know are all massless in the string theory, and differ because they come in representations of symmetries. For example, there is an up quark and a down quark that come together in a doublet of the electroweak theory. When they vibrate the energy of the vibrating string is very large, close to the Planck mass, so it does not affect our world very much. In the string/M-theories all the Standard Model fundamental particles (quarks, leptons, W and Z bosons) are massless. Then they obtain mass because of the Higgs mechanism (which works so that photons and gluons stay massless). The Higgs mechanism works in such a way so that the *theoretical* properties that implied massless particles remain in place, but the *solutions* are allowed to violate it. This is called spontaneous symmetry breaking.

The basic point that **theories** have **solutions**, whose properties can be very different from the properties of the theories, can be explained simply and generally. Suppose a theory is stated in terms of an equation, $X \times Y = 16$. For simplicity consider only positive integer values of X, Y as solutions, and assume that interchanging X and Y gives the same solution. Then there are three solutions, $X = 1$ and $Y = 16$, $X = 2$ and $Y = 8$, and $X = Y = 4$. What is important is that the theory ($XY = 16$) is symmetric if we interchange X and Y, but some solutions are not. The most famous example is that the theory of the Solar System has the Sun at the center and is spherically symmetric, but the planetary orbits are ellipses, not symmetric. The spherical symmetry of the theory misled people to expect circular orbits for centuries. Whenever a symmetric theory has non-symmetric solutions, which is common, it is called spontaneous symmetry breaking.

In this example above, as often in nature, there are several solutions so more information is needed, either theoretical or experimental, to determine nature's solution. We could measure one of X or Y and then the other is determined. Improving the theory leads to an interesting case. Suppose there is an additional theory equation, $X + Y = 10$, also symmetric if we interchange X and Y so the theory remains symmetric. But now there is a **unique solution**, $X = 2$, $Y = 8$, and it is ***not symmetric***. In fact, there are no symmetric solutions. Via the Higgs mechanism all the massless fundamental states except photons and gluons are allowed to be massive even though the theory invariances remain valid.

String theory does not yet have a rigorous definition. Sometimes people state that as a problem or criticism, and some suggest that is a reason not to take string theory seriously. They are unaware or have forgotten that historically the development of theories has always been haphazard. Some results are obtained, and after a while they are understood better, and finally formalized. That is true of Newton's laws, where solving the 'action-at-a-distance' issue took two centuries. It is true of quantum theory, where its formalization started to emerge within a decade of the initial successes, and is still ongoing research. It is true of evolution where the Mendelian genetics underlying heredity were unknown when Darwin wrote. This sort of criticism of string theory is basically irrelevant to understanding our world.

Today it is possible for the first time to address all the basic questions about the laws of nature and the Universe that we see, and its history scientifically, in one comprehensive theory:

- The Planck scales are **exciting** because they provide a natural place to formulate the underlying law(s) of nature.
- Inflation is **exciting** because it explains how the Universe can be old and large and cold and dark, and galaxies.
- The Standard Model is **exciting** because it summarizes four centuries of physics and tells us how our world works.
- The Higgs boson is **exciting** because it completes the Standard Model, and helps us understand quark and lepton and gauge boson masses, and points us toward the supersymmetric world.
- Supersymmetry is **exciting** because it provides the opportunity to combine the Standard Model and string/M-theory and have a window to the Planck scale, without a Hierarchy problem.
- Grand Unified Theories (GUTs) are **exciting** because they tell us how the forces can be unified, and allow us to calculate the electroweak mixing angle.
- Extra dimensions are **exciting** because they allow a UV completion, and the properties of the Planck size region contains much of the physics.
- Compactified string theories and compactified M-theory are **exciting** because they provide a framework that addresses how to explain the Standard Model particles and forces and their properties, and to connect them with gravity in the framework of relativity and quantum theory, and to understand the extra dimensions. As we will see, they also can provide an explanation for having three families.

Being exciting does not guarantee that nature will behave that way, but we should commit ourselves to finding out if it indeed does behave that way, which means we need to build another collider, and we should take the theories more seriously.

References

[1] Margolis H 2002 *It Started with Copernicus* (New York: McGraw-Hill)
[2] Weinberg S 1992 *Dreams of a Final Theory* (New York: Vintage) See also Ooguri H 2004 *Dreams of a Final Theory* YouTube

IOP Publishing

String Theory and the Real World (Second Edition)
The visible sector
Gordon Kane

Chapter 2

The Planck scale—compactification—extra dimensions

The essential logic of our approach is that wanting a *quantum* theory of gravity, a **'UV' complete quantum theory**, implies we should begin with the 10D/11D universe of string/M-theory. Since we live in four dimensions we must then project the higher dimensional universe onto the 4D one, that is, compactify. Note we are inputting data here, the fact that our world apparently is 4D. **Some compactifications, such as M-theory compactified on a 7D 'G_2 manifold',** *automatically* **give the 4D world to be a supersymmetric relativistic quantum field theory, and have Yang–Mills forces and massless quarks and leptons**. We can do a number of tests comparing with our world.

Interestingly, it could be that all compactifications would turn out to give the same theory of the 4D world, but at present it seems like different compactifications give different resulting worlds. Then it would be a great challenge to figure out why one vacuum was deeper than another. People could look for approaches to a comprehensive theory different from the string/M-theory/compactification one, but we will not pursue any other approach here. In terms of explanatory power, the compactified M-theory has been somewhat more successful than others so far, and it is the approach we follow in this book. We encourage string theorists and phenomenologists to work out the possible predictions for several of the other compactifications. There are other features that have to be input, and therefore eventually explained. One major one is that we compactify to small extra dimensions, of Planck-scale size. The present chapter focuses on the definition and meaning of the Planck scales, and on the ways to compactify. The Planck scales for size and time are small, but not smaller than we can comprehend. The next few pages are paraphrased from a similar short section on the Planck scales in my earlier book [1].

Whenever we describe a segment of nature we may have to talk about the actual quantities that are calculated or predicted or explained in units—meters, or seconds, or kilograms or other appropriate units. People have been thinking for well over a century about how to set standard lengths. James Clerk Maxwell, the discoverer of Maxwell's equations of electromagnetism, argued in 1891 that the properties of atoms and of light could be used, and that '… such a standard would be independent of any changes in the dimensions of the Earth, and should be adopted by those who would expect their writings to be more permanent than that body'.

For every theory there is a natural system of units, one in which phenomena are described simply. It would not be sensible to measure room sizes in Planck lengths even though an underlying theory is best written in those units.

Let us illustrate this by considering the units used to describe atoms. The radius of an atom can be computed using quantum theory, and it depends on the mass m, and electric charge e, of an electron, and on Planck's constant h (that sets the scale of all quanta). h is the fundamental, universal constant of quantum theory. Call the electric charge of the electron e and the mass of the electron m. Then in an introductory course in quantum theory one can derive that the simplest atom, hydrogen, has a radius R given by $R = h^2/e^2m$.

This is an astonishing formula. These fundamental inputs determine the size of the atom, even though none of them knows about atoms! Nothing else matters. The nucleus, for example, has a size, but it is just a tiny object at the center of the atom, and its size is irrelevant to the size of the atom. Once we know R, we can express the sizes of all atoms in terms of R, without knowing about h or m or e any more. R is the natural size unit for atoms. Different atoms from the periodic table with different numbers of electrons will have somewhat different sizes, with radii such as $1.2R$, or $2.3R$, but all their sizes will be about R. We hope e and m can be calculated accurately someday in string/M-theory, but e and m cannot be calculated or understood within atomic physics.

There are stronger implications. Given the basic quantities (Planck's constant and the mass and charge of the electron), *the size of all atoms of all kinds, anywhere in the Universe, is determined.* Since mountains and plants and animals are all made of atoms, their sizes are approximately determined by the size of atoms and the electromagnetic and gravitational forces. Since the atoms cannot be made smaller than the size determined by the radius R, having a brain with enough neurons to make enough connections to make decisions about the world requires a minimum size brain. All intelligent life is expected to be about our size, not much larger or much smaller. Nothing the size of a butterfly will be able to think anywhere in the Universe. Thinking organisms could be much larger than people, but they do not need to be.

Suppose a compactified string/M-theory seems to be a candidate for an under-lying theory. We have to express the predictions and explanations in appropriate units. We expect the natural units for the final theory to be very universal ones, not dependent on whether the Universe has people or stars.

There are three and only three universal constants in nature. They are Planck's constant h, the speed of light (denoted by c) that is constant under all conditions, and Newton's constant G that measures the strength of the gravitational force. Einstein proved that energy and mass are convertible into one another. Gravitation is a force proportional to the amount of energy a system has, so everything in the Universe feels the gravitational force. In fact, using these three quantities h, c, G it is possible to construct combinations that have the units of length, time, and energy, and of everything else that might occur in equations. We expect all the quantities that enter into the final theory, or are solutions of the equations of the final theory, to be expressible in terms of the units constructed from h, c, G. Their precise values will not be relevant for the reader.

We expect the electron mass m and the electron charge to also be calculable in terms of h, G, and c in a compactified M-theory, so the atom radius could be expressed in terms of h, G, and c if we wanted. Since that result is complicated we don't exhibit it here.

The Planck length is $(Gh/c^3)^{1/2}$, which is about equal to 10^{-35} meters as we said before, very small, **the natural size of a universe.** The Planck time is $(hG/c^5)^{1/2}$, which is about 10^{-44} sec, **the natural lifetime of a universe, and the Planck mass is $(h.c./G)^{1/2}$, which is about 10^{-8} kg or about 10^{19} GeV** (the beams at the LHC are about 7000 GeV). The Planck mass (or equivalently energy) is very large for a particle.

The physicist Max Planck fully understood over a century ago the universality of those units we call the Planck units. He wrote in his book [2] *The Theory of Heat Radiation* a beautiful and profound idea. I quote it here even though it is a little long. 'All the systems of units which have hitherto been employed…owe their origin to the coincidence of accidental circumstances… In contrast with this it might be of interest to note that…we have the means of establishing units of length, mass, time…which are independent of special bodies or substances, which necessarily retain their significance for all times and for all environments, terrestrial and human or otherwise, and which may, therefore, be described as 'natural units'. The means of determining the units of length, mass, and time…are given by the constant h, together with the magnitude of the velocity of propagation of light in a vacuum, c, and that of the constant of gravitation, G. These quantities retain their natural significance as long as the law of gravitation and that of the propagation of light in a vacuum [and quantum theory] remain valid. They therefore must be found always the same, when measured by the most widely differing intelligences according to the most widely differing methods.' (I have left out some words to make this read smoothly, replacing them with ellipses, and since Planck did not then know about the development of quantum theory I have added that term in square brackets as he would presumably have included it.)

The Planck length and time can also be interpreted as the smallest length and time that we can make sense of in a world described by quantum theory and having a universal gravitational force. To explain that, recall the definition of a black hole. Imagine being on a planet and launching a rocket. If you give the rocket enough speed it can escape the gravitational attraction of the planet and travel into outer space. If the planet has a larger mass, you have to increase the speed needed to escape. If you increase the planet mass so much that the required speed is larger than the speed of light, then the rocket cannot escape because nothing can go faster than light. The rocket, and everything else, is trapped. Light also feels gravitational forces, so beams of light are trapped too. Since gravitational forces grow if you are closer to the center of the planet, when you pack some mass into a sphere of smaller radius it is harder to escape from it. The condition for having a black hole depends on both the amount of mass and the size of the massive object.

Remarkably, if we put an object having the Planck energy or mass in a region with a radius of the Planck length we satisfy the conditions to have a black hole! We cannot separate such a region into parts, or obtain information from a measurement, so we cannot define space to a greater precision than the Planck length! Since distance is speed times time, and speed can be at most the speed of light, and there is a minimum distance we can define, there is also a minimum time we can define, which turns out to be the Planck time.

We saw above that the Planck scale provides the natural units for expressing the final theory when the units are constructed from the fundamental constants h, c, and G. Now we see a second reason for expecting the Planck scale to be the distance and time scales for the final theory—there is no way even in principle to make sense of smaller distances or times. The times when events occur cannot be specified, or even ordered, more precisely than the Planck time.

[There is a third interesting argument that gives the same answer and strengthens the arguments for the Planck scales as the fundamental ones. The gravitational force between two objects is proportional to their energies, and increases as they get closer to each other. Consider for example two protons. They feel a repulsive electrical force and an attractive gravitational force. Normally the electrical force is stronger. But if the energies of the protons are increased to the Planck energy, then the gravitational force between them becomes about equal to the electrical force between them. At the Planck scale all the forces become about the same strength, rather than being widely different in strength as they are in our everyday world. Thus it is reasonable to expect the gravitational force to unify with the others at the Planck scale, as one might hope for in an underlying theory. If one calculates the Standard Model forces as they approach the Planck scale they head the right way to come together, but do not make it very closely. Another argument for super-symmetry is that if one repeats the calculation in the supersymmetric Standard Model the forces do come together almost exactly.]

Finally, then, we understand why the natural size region for compactified string/ M-theories is a Planck-scale size. That is the natural size of a universe—what needs explaining is why some universes (for example, ours) would have three dimensions that have expanded without limit. Similarly, a typical universe should live a Planck time, about 10^{-43} sec. Any universe lasting much longer than that very short time needs explaining. There has been some research studying how theories might imply three large space dimensions, with some good ideas but not yet any compelling ones.

There is a so-called 'anthropic' argument for three large dimensions: it can be shown that planetary and atomic orbits are unstable in any world with more than three space dimensions so these solutions quickly collapse. That was first pointed out over a century ago by Paul Ehrenfest in 1917. Worlds with more than three large space dimensions will not support life. Worlds with fewer than three space dimensions will not have the complexity needed for systems at any level—imagine two lines in a plane trying to cross. They cannot since one cannot leave the plane to get over the other. Perhaps the anthropic argument is indeed the reason for long-lived universes having three space dimensions, but physicists would like to have a more technical, derivable reason, and research will continue toward that goal.

The properties of the Planck-scale size curled up region will contain much information about the original 10D or 11D theory and its predictions. Mathematicians and physicists have studied such regions, particularly the 6D regions from the 10D string theories, in considerable detail, in the theory of manifolds. The 6D regions are called Calabi–Yau manifolds. They have received much attention because the mathematician Shing-Tung Yau proved an important theorem that led to elucidating many relevant properties

[For the compactified M-theories the curled up region is 7D, and lacks a property analogous to Yau's theorem. The 7D manifolds are called G_2 manifolds since they have a mathematical property called holonomy given by the Lie group G_2. For Calabi–Yau manifolds the holonomy group is SU(3). None of these mathematical properties will be relevant for the reader—I mention them so the reader is aware that such features play a role in encoding the full information of the string/M-theory in the curled up region, and that mathematical techniques are being developed to learn the connections to the underlying physics.]

[Although the G_2 manifolds from M-theory compactifications have lagged behind the Calabi–Yau manifolds so far, recently the Simons Foundation has given a large (nearly \$10 M) grant for four years to support studies of 'manifolds of special holonomy', which in practice are mainly G_2 manifolds. Hopefully, soon a number of results important for M-theory compactifications will emerge. The grant has just been renewed for three more years. My distinguished collaborator Bobby Acharya is one of the leaders of the Simons Foundation study.]

Today M-theory and several other versions of string theories are known. No principle has yet emerged that tells us which to compactify to describe our world. People have proceeded appropriately by studying them one at a time and looking at the explanations and predictions that emerge. We will do that for M-theory, where results are very encouraging.

We have seen why the Planck scales are the natural ones. But **our world has a number of other scales we need to understand, such as the electroweak scale where we live (the value of the Higgs field in the vacuum, about 240 GeV) is also a good measure, and the proton mass scale (about 1 GeV). We need to learn why we do not live at the Planck scale.** We will see that compactified string/M-theories naturally generate scales below the Planck energy scale, and large compared to the Planck size and time scales, and provide explanations for the scales.

References

[1] Kane G 2013 *Supersymmetry and Beyond* (New York: Basic Books)
[2] Planck M 1991 *The Theory of Heat Radiation* (New York: Dover)

IOP Publishing

String Theory and the Real World (Second Edition)
The visible sector
Gordon Kane

Chapter 3

Higgs physics: the Hierarchy problem

One very important scale is the one at which supersymmetry is broken. **In physics broken symmetries do not just break up into pieces to throw away.** The original symmetry made several predictions. **The broken one preserves most of those**, and allows one or a few properties to vary from the unbroken symmetry prediction. In the case of supersymmetry, the superpartners continue to exist, one for each particle, with all the expected properties except mass, which can vary. The way super-symmetry is broken sets the scale for all the superpartner masses, so it is very important in terms of searching for superpartners experimentally, and for their phenomenological impact.

We may have to give values for masses in some units. It is probably most clear to consider all scales in terms of masses of particles, or energies (which are conceptually equivalent because of Einstein's $E = mc^2$ relating mass and energy). We can just use GeV units for all masses and scales. Another scale is that of protons and neutrons, which have masses of about 1 GeV. The Planck scale in these units is about 10^{19} GeV. [An electron accelerated by an electrical potential difference of one volt will gain an energy of 1 eV. The prefix G stands for a billion (10^9), and T for 10^{12}. 1 TeV is 1000 GeV. These are units we will rarely use.] We need to understand this large ratio between the Planck scale and the electroweak scale (10^{17}) or the proton mass scale (10^{19}).

The theory of the Standard Model implies that the quark and electron masses, and also the masses of the W and Z bosons that mediate the weak interactions, are zero for the unbroken electroweak symmetry. Those particles of course actually have non-zero mass in the real world, and the theory allows them to have their correct masses by violating the electroweak symmetry in a very special way, called the Higgs mechanism.

The Higgs mechanism arises from properties of the Higgs field, and the Higgs bosons discovered in 2012 at the European CERN Laboratory in Geneva, Switzerland, are quanta of the Higgs field. Finding the Higgs boson (with its observed properties) implies that the Higgs field exists, a field that pervades the Universe. For most purposes one can think of the Higgs field like other fields in nature, for example, gravitational fields, or electromagnetic fields.

But there are two surprising features of the Higgs field. Most fields arise from charges. Electromagnetic fields arise from electrical charges, gravitational fields from mass and energy, and so on. The Higgs field does not have sources, it is effectively just there. Second, other fields vanish in the absence of charges, but the Higgs field is not zero in the vacuum. In the original formulation of the Standard Model these two features were assumptions, made by clever theorists after years of efforts. In 1982 it was realized that in a supersymmetric world **these features could be derived instead of assumed**, which was a significant advance, and significant evidence for supersymmetry. A necessary condition for the derivation was the existence of a heavy quark (heavier than the W boson), presumably the top quark since all the other quarks had been observed and were not heavy. The top quark was eventually observed at Fermilab in the late 1990s and was indeed heavy, over twice the mass of the W boson. This was a successful prediction of the supersymmetric theory.

One would expect that the ground state of the Universe, the 'vacuum' state, is the state with the lowest energy, since any system will eventually end in the state with lowest energy. One would think that state would be the one where all fields are zero, since fields carry energy. Remarkably, in our Universe the state with the Higgs field being non-zero has a lower energy than if the Higgs field were zero. We learn the Higgs field is non-zero in the vacuum from implications of the data on how Higgs bosons decay. If the electroweak symmetry were exact, the Higgs boson would be forbidden to have the decays h → W^+W^- or ZZ, but in fact those decays occur in the LHC data and have a large rate. The presence of this decay demonstrates that the Higgs field is non-zero in the vacuum. These decays are not suppressed, so the Higgs field vacuum value is full strength. [The technical derivation is in brackets below and can be ignored by most readers.]

[Particles have a property called weak isospin, analogous to spin mathematically. The W and Z both have weak isospin of one. The Higgs boson has weak isospin ½. Two weak isospin 1 particles cannot be combined to give a weak isospin ½ particle. The true interaction Lagrangian must be ± of the form HHWW, which is allowed since the Higgs pair can have integer weak isospin. Then one Higgs field H takes on its vacuum value and the effective Lagrangian is proportional to HWW, leading to the above decay. The ZZ rate is somewhat more accurately measured so far, and has about a 10% uncertainty, so most likely the observed Higgs boson has the full Higgs vacuum expectation value but there is currently still room for some of the other bosons of the Higgs field to carry a small vacuum value. As more data comes from the upgraded LHC the ratio of the production rate times the decay to the Standard Model predictions should converge on a value of 1 with a small error such as ±0.01.

The Higgs boson decay branching ratios to fermions depend on the final state fermion mass with a factor of mass cubed. Sometimes that is taken as evidence for Higgs physics, but it actually holds for decays of any scalar particle, even if it has a vanishing vacuum value so it is only evidence for scalars.]

Now we turn to understanding the Higgs mass value. The ratio of the electroweak scale to the Planck scale is a tiny number, and understanding that ratio is one of the main challenges in extending the Standard Model, and in going beyond the Standard Model toward an underlying theory, perhaps the most difficult problem, the 'Hierarchy problem'. It arises because our world is described by relativistic quantum field theories. Then the masses of particles can change since there is some probability that any particle can fluctuate into any other (via 'virtual particles'). For most properties the change in value is very small, but for the mass of the Higgs boson it turns out for technical reasons to be unlimited—when the full effects of virtual particles are included the Higgs boson mass becomes as large as it can be, which is the Planck mass.

Interestingly, **in compactified models such as the compactified M-theory the Higgs boson mass can be calculated, and comes out correct. That calculation was done before the CERN discovery of the Higgs boson and constitutes 'a prediction from string theory'** and 'a correct prediction from string (really M-theory)', contrary to what lots of string theorists and reporters say. There will be more predictions in the following.

An amusing feature of the prediction is that the theory calculation could never be done with an accuracy better than (about) 1% given the inputs needed and the intrinsic errors of the calculations. The experimental measurement is already a lot more accurate, about 0.1%. Improving the accuracy of the Higgs boson mass measurement will not be interesting since it cannot be calculated to better than the present measurement accuracy.

The Higgs mechanism that solves the problem of the masses of quarks and electrons and W, Z bosons does so by giving all of them mass proportional to the Higgs boson mass, so the Hierarchy problem pushes all their masses to near the Planck scale rather than the actual values at the electroweak scale and smaller. That is why it is a serious problem, although it is a conceptual problem in practice. In the Standard Model one ignores the prediction that the masses are near the Planck scale and assigns them their measured value.

In the Standard Model itself that problem is unavoidable. It is a theoretical problem, so it can be ignored in practice, but we hope for a theory without such apparent inconsistencies. One way the Hierarchy problem might be solved is by extending the Standard Model to be a supersymmetric theory. That means that every particle of the Standard Model (i.e. the quarks and leptons, photons, W and Z, gluons) has to have a partner that differs by half a unit of spin and perhaps differs in mass. The partners must exist. They might be detectable at the LHC. For M-theory compactified on the G_2 manifold the resulting 4D theory must be (a broken) supersymmetric one. That helped convince me and my collaborators to focus on that compactification. Having to double the number of particles may seem uncomfortable, but the same thing happened when charge conjugation invariance led to a doubling of the number of particles for antiparticles.

This solves the Hierarchy problem because in quantum field theory the virtual particle contributions from pairs of particles differing by half a unit of spin have opposite signs, so those from each particle and its superpartner just cancel if they have the same mass. When they have different masses the cancellation is incomplete, and leave us with a 'little Hierarchy problem' that is an interesting real problem but far less of an issue than the full Hierarchy problem.

Supersymmetry is another hidden (so far) aspect of nature. The world needs to have the whole set of superpartner particles that have not yet been experimentally observed. They have been given nice names: photon and photino, W and wino, electron and selectron, quarks and squarks, and so on. It is clear that electron superpartners with the same mass as the electron do not exist or they would have bound with anti-protons to make new kinds of 'atoms' very different from ordinary atoms, and they would already have been observed. Lots of phenomena would have occurred that did not. It is also clear in some compactified string/M-theories that the supersymmetry is inevitably present in the theory, and is indeed a broken symmetry, with the superpartners expected to be heavier than the Standard Model particles. The symmetry is broken by the theory itself, not by some external mechanism. In particular, that is true in the compactified M-theory described later.

The Standard Model is essentially a complete theory in its domain, so it can be used to calculate observables for experiments. But the supersymmetric extension of the Standard Model is not yet a complete theory because of the broken super-symmetry, so the expected masses of the superpartners cannot yet be generally predicted. **A theory with unbroken supersymmetry would actually introduce no new parameters into the supersymmetric Standard Model.** None. If we make assumptions about how supersymmetry is broken, we can calculate observables. One argument is that 'naturalness' can be used to estimate what values some observables might have. Basically naturalness was the point of view that superpartners should be the same as their Standard Model partners except for masses, and the masses were not very large if they were to solve the problems they were good for. If those arguments were correct some superpartners would already have been discovered at the CERN LHC. It would have been nice if the naturalness arguments had worked, but they did not. Since they were not predictions from a theory it is not clear how to interpret that.

In principle one can carry out compactifications, which generally lead to super-symmetry breaking, and calculate superpartner masses for each compactification. Technically the calculations are difficult, so only a few have been studied. Consider the gluino, partner of the gluon. It has strong QCD interactions just as the gluon does, so it should be one of the earliest particles produced at the LHC. The compactifications studied so far suggest the gluino mass could be as light as about 1500 GeV, but not much lighter, in which case it should be observable in the current three-year LHC run. Observing it is not straightforward (see the supersymmetry section later). It probably would have been observed at 1500 GeV or lighter but not at 1700 GeV, which is within the uncertainty range of the theory.

More generally, the LHC is now finally operating in the region of energy and intensity where the compactified theories suggest it is reasonable to be optimistic about superpartner discovery, gluinos in particular. Later we will look in more detail at the superpartner spectrum predicted by the compactified M-theory. Other string theory branches seem to give different predictions for the spectrum, perhaps with somewhat heavier gluinos. We will see below that the compactified M-theory predicts that four superpartners can be discovered at the LHC, gluinos, and charged and neutral winos (gluinos, eight treated as one particle since their color is not directly observable; charginos; neutralinos: LSP). The lightest superpartner emerges in gluino and wino decays, so its presence can be inferred. In the compactified M-theory, all other superpartners are too heavy for the LHC until it is upgraded in luminosity and/or energy, probably too heavy even then.

Exciting implications of the existence and properties of the observed Higgs boson

Nature has played an amusing and challenging trick on us. The Higgs boson discovered at LHC seems to be a Standard Model one, if one looks at its decays or for Higgs partners. But it cannot be a Standard Model one because of the Hierarchy problem. Fortunately, there is a well-known supersymmetric model whose Higgs sector contains a Higgs field whose Higgs boson looks a lot like the Standard Model one. It did not have to come out that way. There could have been several Higgs field bosons sharing the vacuum value.

For anyone who cares what the Higgs boson is telling us, there are four good clues.

1. **In the minimal supersymmetric Standard Model (with a pair of decoupling Higgs doublets) there is an upper limit on the Higgs boson mass of about 130 GeV.** That's satisfied for the observed Higgs boson mass. There is a tree level upper limit of M_Z [91 GeV], and quantum corrections can raise it up to about 130 GeV. These are larger than usual due to the large top quark Yukawa coupling. The observed mass satisfied this condition, so interpreting it as the lightest Higgs sector state of a two-doublet decoupling Higgs sector is natural.

2. **In a world with electroweak scale superpartners the Hierarchy problem is solved.**

3. **There is a well-known model, the two-doublet Higgs sector, satisfying the electroweak symmetry breaking conditions.** It has one light Higgs eigenstate and four heavy ones. It's decay branching ratios for the lightest one are very close to the Standard Model ones. It's called the **decoupling solution** (mentioned in [1]) and has been known for two decades. As the name suggests, the Higgs states besides the lightest one get rather heavy [they will be approximately equal to the gravitino mass, a few tens of TeV, which we'll describe a little later. The decoupling solution arises naturally in the UV complete theories such as compactified M-theory.]

4. **The fourth clue is more subtle. For a single Standard Model Higgs boson the potential could become negative, and also could become large in magnitude, an unacceptable physical situation.** We describe it as 'unbounded from below'. Then there is no stable world. Some people have noticed that the Universe can be in a stable state such as ours, and reach the unstable one by decaying after a time much longer than our Universe has existed. **But then people realized that during inflation the Higgs field fluctuates a lot, and would often push the Universe into a negative potential state.**

Such a state is not allowed. Inflation, and the Big Bang would be unlikely to have happened. It looks like this instability is real, and some new physics must occur to stabilize the vacuum.

In a supersymmetric world the coefficient of the quartic term is determined by the electroweak gauge couplings, and is positive definite.

It's remarkable that the phenomenological two-Higgs doublet solution exists and that it accommodated points 1–4. These features are very encouraging from the point of view of finding the underlying theory.

IOP Publishing

String Theory and the Real World (Second Edition)
The visible sector
Gordon Kane

Chapter 4

Supersymmetry

We've mentioned supersymmetry before. Now we describe the supersymmetry idea further, in some detail, and emphasize what supersymmetry adds.

Supersymmetry is the surprising idea, or hypothesis, that at the deepest level, for the ultimate or final theory, the laws of nature don't change if fermions (particles with half-integer spin) are transformed into bosons (particles with integer spin, including zero), and vice versa. If nature is indeed supersymmetric, the *equations* of the final theory will remain unchanged even if fermions are replaced by bosons and vice versa, in an appropriate way. This should be so in spite of the apparent differences between how bosons and fermions are treated in the Standard Model and in quantum theory. The details of those differences don't matter for us.

Whenever an extension of the existing description of nature is proposed it must pass many obstacles before it has a chance of being valid. It must be consistent with the rules of quantum theory and special relativity, and it must not change any of the tested consequences of the Standard Model. The fermionic or bosonic nature of particles comes from their spin, and spin is related to quantum theory and special relativity, both of which in turn involve space and time in their formulation. Thus the formulation of supersymmetry must also involve space and time as well as interchange of bosons and fermions. It is remarkable that such a property can be introduced without coming into conflict with some already established result.

The reader may nevertheless be underwhelmed. Who cares if the final theory is invariant when its fermions and bosons are interchanged? OK, it's remarkable that can be done in a relativistically invariant quantum theory, but so what? Actually, that's the way physicists originally felt about it too, for the most part. The supersymmetric theory was technically very beautiful, written in the mid-1970s, so theorists enthusiastically studied it, but it was spoken of as a 'solution in search of a problem'. I got into the field because I asked at a number of talks in the late 1970s how we would know if nature were actually supersymmetric or not, and the typical answer was that no one had thought very much about that. It was such a beautiful theory that we began to feel strongly that it was important to figure out how to test it experimentally.

There is indeed reason to be excited, but it wasn't obvious at the beginning. That's related to how supersymmetry arose as an idea, in a manner different from other ideas in the history of science. New ideas, including the Standard Model, had always come as a response to trying to understand observed regularities, or to puzzles, to a desire to explain the behavior of aspects of the natural world, or to apparent inconsistencies in existing descriptions. Supersymmetry on the contrary was originally noticed as a property of certain models (actually, with fewer than three space dimensions) being studied for their own sake. A fascinating aspect of the history is that supersymmetry was not introduced to solve any experimental mystery or theoretical inconsistency. Over a decade as it was studied and better understood, theorists realized that it actually could solve a number of the important mysteries in particle physics, and provide new approaches to others. For many physicists, that supersymmetry solved problems it was not introduced to solve was itself a powerful hint that it was indeed part of the description of nature. No other major physics concept has ever been introduced as an idea not related to any data or inconsistency, and then found later to solve important problems.

We will list a few of these mysteries with minimal explanation to give an overview of the impressive impact of the supersymmetric theory. We have been using the word 'mystery' in the detective story sense. We expect mysteries to have solutions, to be solved—usually by the end of the book. The following list is to indicate why many physicists work on the theory of supersymmetry or on testing it experimentally, and why there are over 20 000 papers on supersymmetry before there is explicit evidence that it is really how nature works.

In order to incorporate the masses of the particles into the Standard Model description, physicists postulated the existence of a Higgs field. They also assumed it interacted in a very specific and somewhat enigmatic way. The Higgs physics was a mysterious part of the Standard Model, hard to accept and hard to test, though technically it solved the problem for which it was introduced. In particular, some new physics beyond the Standard Model *must* exist to provide the Higgs physics—the Standard Model simply cannot provide an explanation of the Higgs interaction needed to account for the masses of particles in a consistent way. Then in 1982 several theorists figured out that if the Standard Model was extended to be supersymmetric it could provide an elegant physical explanation for the Higgs physics. For many theoreticians that was the result that convinced us that supersymmetry was a property of nature, not just nice mathematics. Even better, in order for the supersymmetric approach to Higgs physics to work it was necessary that the top quark (whose mass was only measured in the 1990s) be unusually heavy compared to the other quarks and leptons—that the top was predicted to be heavy and was indeed confirmed to be heavy a decade later by data was a powerful indirect test of the validity of supersymmetry.

- The Standard Model by itself has a very serious conceptual problem. We saw above that the natural scale for the final theory was the Planck scale, about 10^{-35} meters. The Standard Model is a description of quarks and leptons and their interactions, at a scale of about 10^{-17} meters. The problem is that, in a quantum theory, physics at every scale may contribute to physics at every other scale, so it may not be consistent to have these two scales so separate—the Standard Model scale and the Planck scale should be very near each other, and in fact, more precisely, the Standard Model scale should be near the Planck scale. Another way to view this problem comes from recognizing that in the Standard Model all the masses of electrons, quarks, Ws and Zs, should either be zero or about the Planck mass. For the Standard Model this is indeed a major problem, even though it is a conceptual problem that does not explicitly impact the experimental predictions of the Standard Model.

This is the Hierarchy problem described earlier. It's the central problem in particle physics today. The problem has two parts. First, given that there is a separation of the Standard Model scale from the Planck scale, why does the Standard Model end up where it is (at about 10^{-17} meters, or about 100 GeV) and not at some other scale? Second, and more important conceptually, what can make the theory maintain that separation in a mathematically consistent way? The supersymmetric Standard Model solves the second problem and gives insight into the first. It does so in a manner that uses the unification of fermions and bosons in an essential way. The very nature of fermions and bosons implies that they contribute to the coming together of scales in ways that cancel, so the mixing of scales can be canceled in a general way and solve this problem.

- For two centuries physicists have been actively trying to unify our description of the forces of nature. Having five different forces, rather than one basic force, suggested we were missing some unifying principles. Maxwell succeeded in unifying electricity and magnetism, and the Standard Model unifies the description of weak interactions with electromagnetism, so there has been some progress.

- In a quantum theory one can calculate how a force would behave if one could study it as smaller distances. Remarkably, when this is done in the Standard Model for the electromagnetic, weak, and strong forces it is found that they become more and more like each other at shorter distances, though finally they do not quite become equal in strength at any distance. More remarkably, when the study is repeated with the supersymmetric Standard Model, as was done in the early 1980s, it is found the forces do become essentially equal at a very small distance, somewhat larger than the Planck scale.

- That did not have to happen—nothing in the Standard Model implies that the forces should become equal. And if they do become equal, nothing in the Standard Model implies that should happen at just the distance scale that would be expected if full unification occurred at the Planck scale. Presumably these are important clues toward a better understanding. Since we do not know the form of the complete theory at such small distances very well, we do not know how to extrapolate to even smaller distances, but some reasonable ways of doing that suggest that these forces become equal to the gravitational force at about the Planck scale, encouraging very much the idea that indeed the goal of understanding the forces of nature in a simple way will be reached. The superpartners seem to be needed for this to work.

- All of the superpartners are expected to be unstable particles, decaying into lighter superpartners, except possibly for the lightest superpartner (LSP), which has no lighter ones to decay into and seems likely to be stable. Thus supersymmetry may introduce a new stable particle into the Universe, joining photons, electrons, neutrinos, and protons. The light we see from stars is composed of photons. Protons and electrons form the stars and planets. Neutrinos, and the LSP (if it exists and is stable), will be forms of matter that are present throughout the Universe. Since neutrinos only feel the weak and gravitational forces, not the electromagnetic or strong ones, they will not participate in forming stars. They will be 'dark matter' (actually about 1% of the observed amount).

Supersymmetry suggests that there may be dark matter composed of the LSP. Right after the Big Bang there were about the same number of each kind of particle. Most particles decayed into lighter ones, and some annihilated into others as the Universe expanded and cooled. Since we have a theory of how they all interacted, we can calculate how many are left now. Even though they did not coalesce into stars that produce photons that we can see, their presence can be detected by their gravitational attraction for what we do see, if there are enough of them—their presence modifies how stars move in galaxies and how galaxies move relative to one another. It was realized in the early 1980s that supersymmetry predicted that there should be considerably more LSP dark matter than even the matter in stars. Astronomers indeed had already observed that the Universe did have considerable dark matter, because stars and galaxies did not move through the Universe as they would if the only matter was what we could see, but at that time it was not known experimentally whether the dark matter could be non-luminous forms of ordinary matter (e.g. interstellar dust), or if a previously unknown kind of matter must exist. We will examine the dark matter and how to study it in the laboratory in chapter 9.

Remarkably, the stability of the LSP looks different if we treat the supersymmetric Standard Model as an effective theory at the electroweak scale from how it looks as a compactified remnant of a high scale string or M-theory. In the latter case one recognizes that the LSP is very unlikely to be stable. It decays to hidden sector matter.

- The LEP collider at CERN and the SLC collider at Stanford were constructed during the 1980s in order to test the Standard Model, and to search for new physics that would strengthen the foundations of the Standard Model. Before LEP and SLC operated, it was possible to predict the kinds of results they should find in a supersymmetric world. Either superpartners would actually be produced and detected, or else the effects of supersymmetry on the quantities measured were expected to be very small, so observables should have essentially the values the Standard Model predicts for them. Non-supersymmetric approaches to an explanation of the Higgs physics generally predicted larger effects. After a decade of accurate measurements, no superpartners were produced, unfortunately, but the results did confirm the prediction that there were no sizeable deviations from the Standard Model expectations. This is a subtle but serious argument. It could easily happen that one or several observables were measured to differ from their Standard Model predictions due to virtual superpartners, yet no super-partners were detected. (If you have sufficient confidence in indirect arguments, you might even conclude that the combination of the need for some extension of the Standard Model to explain the Higgs physics, plus the absence of any deviations from the Standard Model expectations for observables in the LEP and SLC data, together confirm that supersymmetry must be a part of the description of the world. But most physicists want verification that is less indirect.)

- We have already seen that supersymmetry provides two (consistent) ways that suggest the description of the electromagnetic, weak, and strong forces will be unified with gravity. One was the way they became of the same strength at very small distances near the Planck scale. The second was that super-symmetry as a symmetry throughout space and time necessarily had a connection to the theoretical description of gravity (more on this below). While the connection of supersymmetry to gravity is not yet fully understood, it is very encouraging that the supersymmetric Standard Model is related to the theory of gravity, while the Standard Model is not.

- Supersymmetry has led to new approaches to solving or explaining a number of other important problems that had no possible solution in the Standard Model. Here we will just list some to give a sense of the opportunities, without explaining them or any jargon; later we will examine some of them. Fundamental questions for which supersymmetry provides new ideas and methods include understanding how the Universe got to be mainly matter and not antimatter, whether and how protons decay, why the Universe is the age and size it is, rare decays of quarks and leptons, and more.

- **Finally, probably the most important consequence if supersymmetry is part of our description of nature is that it provides a window to look at the Planck scale from our world that is so distant from the Planck scale.** We have seen that there are several reasons to expect that the final theory will be naturally formulated at the Planck scale. But we cannot ever hope to do actual experiments at the Planck scale—it is just too small. **If supersymmetry is part of our description of nature it implies that we can write the predictions of a candidate for the final theory at the Planck scale, and then calculate what should be observed in experiments we can do at colliders, for neutrino masses, for proton decay, and much more. Similarly, we can measure parameters that are part of the description of the theory in experiments and then calculate the values they have at the Planck scale (many of them have values that depend on the distance scale, which is common in quantum theory).** Most likely the latter procedure will be the actual one, with experimental input needed before it is possible to formulate detailed candidates for the final theory—we can hope someone is smart enough to guess the final theory, but if not we can get there anyhow by building up knowledge about the form the theory must take, using experiment and theory together as physicists have traditionally done so well.

- Even if someone does guess the final theory, they won't be sure, and no one will believe it, unless there is experimental input of the kind supersymmetry can provide. **If the world is not supersymmetric we do not know of any way to relate physics at the Planck scale to physics at our scale. If the world is supersymmetric we can expect both to be able to formulate the final theory and to test it.**

The above arguments were of two kinds. Some provided explanations for actual phenomena, or predicted observations, while others were about our opportunities to make sense of the world at the deepest levels. The first kind provides real evidence that the world is indeed supersymmetric, indirect but still real and significant evidence. The second kind, of course, does not imply that supersymmetry is real— just because it will help us solve String/M-theories or provide a window on the Planck scale does not mean nature is that way. But both kinds of reasons contribute to the very strong and widespread interest in supersymmetry among physicists. And it is worth repeating that supersymmetry was not invented for any of these reasons— all the reasons emerged as its implications were studied. Indeed, the results described above provide impressive examples of emergent properties of a basic theory, properties implied in a sense by the theory but not apparent until suggested by data or after much study.

Supersymmetry has the consequences it does because the requirement that the theory is unchanged when bosons and fermions are interchanged can only be satisfied if the theory (the equations) is constrained to have a certain form. It may be helpful to mention a couple of historical examples of properties that emerged as theories became more constrained. Maxwell took several equations people had developed over decades to explain electrical and magnetic phenomena and tried to combine them into one consistent set. He found that to do that he had to add a term to one equation. The new set of equations then turned out to have a new solution that described light as an electromagnetic wave—suddenly the behavior of light was incorporated into the electromagnetic theory, and electromagnetic waves we now use for communication were unexpectedly predicted to exist. Another example comes from special relativity. Einstein found that to have the same laws of electromagnetism hold in all systems shifted in position or speed relative to one another it was unavoidable that space and time got tied together. In any mathematical science, new constraints on the basic equations often imply major new physical phenomena, just as we have seen in the case of supersymmetry.

In chapter 1 we saw that an important aspect of the Standard Model was the invariance of the theory under the interchange of certain particles, the electron and the electron neutrino, the up and down quarks, and so on. Some of the particles required for that invariance to be valid already were known to exist when the Standard Model was developed, but others were later predicted to exist and then they were found. We also saw that having the theory be consistent both with quantum theory and special relativity required the existence of antiparticles. None of the antiparticles had been observed at the time they were predicted, and because physicists had learned what to look for, all have been observed since then. For supersymmetry to be valid it would again be necessary for previously unknown particles to exist that are the same as the particles of the Standard Model, except that bosons ↔ fermions. Since bosons and fermions have different spins, the partners must differ in their spins. There must be another particle just like a photon, with no electric charge or weak charge or strong charge, but with spin one-half instead of spin one. There must be another particle just like the electron, with the same electric charge and weak charge as the electron, no strong charge, but with spin zero instead of spin one-half. An examination of the particles we know shows that none of them have the properties to be the needed partners. The situation is analogous to what happened with antiparticles, where all the predicted new particles had yet to be found. We call the new set of particles 'superpartners', or 'sparticles'. The sparticles form smatter.

As our understanding of nature progressed through the past century we saw the number of 'electrons' grow, in a sense. From 1895 till the 1920s there was thought to be only a single electron. Then first the data about energy levels of atoms, and later Dirac's unification of quantum theory and special relativity led to assigning half a unit of spin to the electron. In the quantum theory that meant the electron spin could point up or down with respect to any arbitrary direction, and accounted for the observed extra energy levels of atoms. It is as if one should think of two electrons, one with spin up and one with spin down. Then antiparticles were predicted and found. Again, it is as if electrons had to come in four kinds, particle or antiparticle each with spin up or spin down. With the validation of the Standard Model in the 1970s we learned that electrons and neutrinos were really different projections of the same basic particle that could turn into each other by emission or absorption of a W boson, with analogous processes for antielectrons and antineutrinos, so in a sense we think of one object that can exist with spin up or down, and as electron or antielectron, neutrino or antineutrino. With supersymmetry we go one more step. Each of these states of the 'electron' is predicted to have a partner that differs only in its spin being zero rather than one-half. **Finally, there are 16 electrons.**

This might seem to be a proliferation of particles. But if **the theory says that given any one of them all the others must exist**, then the apparent proliferation follows from a conceptually simple structure. Just as a simple structure, based on the electron and the up and down quarks and their interactions, underlies the great complexity of our world, so there is a simple conceptual structure with a small set of Standard Model particles plus their antiparticles and Standard Model partners and superpartners.

There is a simple notation and terminology for superpartners. The superpartner of every particle is written as the partner with a tilde (~) over it: $e \rightarrow \tilde{e}$, γ(photon) $\rightarrow \tilde{\gamma}$ etc. If the Standard Model particle is a fermion the name of the partner is the fermion name with an s- in front—sfermion, selectron, squark, up squark, stop, sneutrino, etc. If the Standard Model particle is a boson the name of the partner is the boson name with an –ino suffix—photino, gravitino, Wino, higgsino, etc. Since all of the regularities of the Standard Model hold for the supersymmetric Standard

Model too, keeping the names helps in keeping track of processes and predictions. As many of us have said, we have a new slanguage.

It should be emphasized that supersymmetry is the *idea* that the laws of nature are unchanged if fermions ↔ bosons. It is not that an electron becomes a selectron in the equations that constitute the basic theory, but that the equations should contain symbols representing both electrons and selectrons, and the equations are unchanged if those symbols are interchanged. The existence (or not) of the sparticles themselves is the most dramatic prediction to test that idea—they are the anticipated smoking gun. There are several other tests that are less explicit. Some indirect tests based on the implications of supersymmetry have already been positive for super-symmetry, helping convince many physicists that it is indeed correct—those are the ones listed earlier in the chapter, and we will examine some in detail in later chapters. Supersymmetry requires not only new (to us) kinds of particles, but new interactions involving the new particles. The theory implies that the new interactions have related strengths. That's an important prediction, and will be an important test of the theory if some of the selectrons and photinos and other sparticles are observed.

4.1 Supersymmetry as a space–time symmetry—superspace

We have seen that there are two important ways to think of supersymmetry. They are consistent, of course, but often one or the other is most helpful to achieve a given goal. The first and most common way is as a symmetry of the laws of nature under the interchange of bosons and fermions. A sparticle is predicted to exist for each of the quarks and leptons and bosons of the Standard Model. The experimental approach to supersymmetry and most practical calculations are carried out in this framework. A second approach was thinking of supersymmetry as an effective theory that opens a window on the Planck scale, so that we can formulate and test the final theory, and shows us how important supersymmetry is in our search to understand our Universe. There is a third way. Sometimes it is fruitful to think of supersymmetry as a space–time symmetry, but in an extended space–time, called 'superspace'. This approach is often most helpful to uncover the mathematical properties of the supersymmetry theory. The remainder of this section is a little more technical than the rest of the book, and is not needed to follow the rest.

Think about an isolated particle or object, one on which no forces act. Assume it's moving, with some energy and momentum. If there were forces they could change its energy and momentum, so with no forces we expect the energy and momentum to be unchanged. We say the energy and momentum are 'conserved'. We are making assumptions when we assume the energy and momentum are conserved in the absence of forces. We expect space and time to be neutral, so the object can move through them without feeling effects. If no other objects are around, we don't think it matters whether our object is here or a few miles away, or rotated some amount, because we think space and time don't affect objects in them. Whenever something is invariant under changes we speak of a 'symmetry'. Whenever some quantity is conserved there is an associated symmetry, and vice versa. Just as we can change our position in space or time and find symmetries, we can change our position in superspace and find an associated symmetry—supersymmetry.

Symmetries in physics can be of two kinds. The examples we just discussed are 'geometrical' or space–time symmetries. In chapter 2 we saw that the Standard Model has some 'internal' symmetries, ones for which the theory is invariant under interchange of particles such as up and down quarks. A very important question is whether there could ever be a symmetry that mixed geometrical and internal symmetries. For example, we can be pretty sure that moving an up quark a few centimeters won't change it into a down quark, but maybe there is some more subtle geometrical change that could? It turns out that for our standard ideas of space–time the naïve approach is right, and geometrical and internal symmetries cannot be mixed. It can be proved that the space–time symmetries we already know about are the only possible ones. However, if we allow for new 'fermionic' dimensions then it turns out that one more symmetry can exist, and it is supersymmetry. Another attractive feature of supersymmetry is that it is the last possible space–time symmetry nature could have, the only mathematically possible symmetry that is not already known to be part of nature.

Supersymmetry is the idea that the basic laws are invariant under interchanging fermions and bosons. We just saw that invariances lead to symmetries. In our usual space–time bosons and fermions behave differently. It turns out that we can attach to our normal space–time another four 'dimensions' that will allow us to incorporate differences between fermions and bosons into the structure of the 'space', defining a 'superspace'. Superspace is a geometrical structure in which fermions and bosons can be treated fully symmetrically.

The extra superspace dimensions are not like our dimensions. They have no size at all. Nor are they like the extra small dimensions of string/M-theory. In the superspace we can think of bosons and fermions as two different projections of a single object, similarly to the way an electron and its neutrino can be thought of as two different projections of a single object in an internal space. Superspace is an intrinsically quantum theoretical structure, because the whole idea of fermions is meaningful only in a world described by quantum theory. One can think of superspace as adding fermionic dimensions to our usual bosonic dimensions. If the world really exists in superspace it affects observables—it is a testable idea. The simplest and most dramatic test is that the superpartners must exist. If superpartners are found we can interpret the results as letting us learn experimentally about the properties of superspace.

One can turn the history around. Suppose from the beginning quantum theory had been formulated in superspace. Then immediately it would have predicted that all particles would exist in both a bosonic and a fermionic form. Supersymmetry, formulated in superspace, requires that fermions exist—in a sense it explains their existence.

Einstein's general relativity is a geometrical theory of gravity, which views the gravitational force as an effect of the distortion of space–time by masses (e.g. planets). Once superspace was formulated, people immediately thought of using it as the basis of a generalized geometrical theory of gravity, 'supergravity'. Supergravity incorporates general relativity, and extends it. The graviton that mediates gravity is predicted to have a superpartner, the gravitino. The connection of supersymmetry to gravity encourages people to think that the unification of gravity with the Standard Model forces will include supersymmetry. Interestingly, the geometrical structure of supergravity takes a simple and elegant form if the space–time is extended to eleven space–time dimensions (with associated superspace coordinates). As it happens that is also the largest space–time dimension in which one can write a consistent theory of gravity.

4.2 Hidden or 'broken' supersymmetry

So far we have argued that there is good reason to expect that the laws of nature are supersymmetric. In fact, the situation is more subtle. We know that nature is not exactly supersymmetric for two kinds of reasons. First, if the world had a selectron with every property identical to that of the electron except its spin, we would already have observed it in experiments. That's true for a number of other sparticles too. Second, even if the relevant experiments were not possible, we can deduce that the world would be entirely different if selectrons with the same mass as the electron existed. In fact, there would probably be no life in the world. That is because if bosonic electrons existed they would all fall into the lowest energy level of an atom, shielding the electric charge of the nucleus, and there would be no valence electrons to bind atoms into molecules. This reason should not be taken as an excuse for supersymmetry to be broken, but it illustrates how indirect analyses can be helpful in understanding how the world works.

In physics, the symmetries systems have can be powerful guides to how the system behaves. Some symmetries are exact ones, while many are not. Even when they are not exact they still can be very helpful in understanding the behavior of the system. Take a very simple example—think of a child's spinning top. It has a cylindrical symmetry about the axis it spins around. Suppose it were poorly made, somewhat bumpy. It would still spin. Maybe it would wobble a bit if it were a little heavier on one side. It may spin a little less long, and be more likely to end up lying on one side than another, but it is still recognizably and basically a top even though its cylindrical symmetry is not quite right. When a symmetry is only partly valid or perfect, physicists speak of it as a broken or hidden symmetry. An example closer to our interests is the symmetry between particles and antiparticles found by Dirac from which he deduced the existence of antiparticles. That symmetry was called 'charge conjugation invariance' since Dirac's equation did not change when the sign of the charge was reversed. It is a valid symmetry for the particles alone, and also valid for the ways the particles interact via the electromagnetic and strong forces, but it turns out it is not valid for interactions via the weak force. The important point is that despite the partially broken symmetry all the predicted particles (i.e. the antiparticles) have been found to exist. The supersymmetry case is expected to be somewhat similar to the antiparticle one. All of the superpartners should exist, but they can have masses different from their particle partners. The interactions will be similar, but can differ because the masses are different, and for other associated reasons. If the superpartners are heavier, they might be too heavy to have been observed in experiments so far.

The issue of hidden supersymmetry illustrates well how physics progresses. In the 1960s there was no theory of the forces or basic particles at all. Then the Standard Model emerged. It provided a very good description of the particles and interactions. Its remaining problem was to determine the form the Higgs physics takes so that the particle masses could be incorporated into the theory, and to understand from what underlying physics the Higgs physics itself arose. Supersymmetry in turn can provide the answer to the Standard Model Higgs physics problem, i.e. it explains how the Higgs physics arose. The supersymmetry answer actually depends on the way that supersymmetry is hidden or broken, and ties the supersymmetry breaking to the Higgs physics of the Standard Model (more about that in the next paragraph). We do not yet understand how the supersymmetry is broken—that should be explained by the next level of effective theory; probably the supersymmetry breaking won't be explained within the supersymmetric Standard Model itself. But supersymmetry is a sufficiently comprehensive theory so that it is possible to know the form the broken supersymmetry theory must take, even though we do not yet understand the mechanism that generates that form, very much like the way we could incorporate the masses using the Higgs physics even though we did not yet understand how the Higgs physics arose. A shortsighted view might claim that we had just traded one problem for another, but that is really not so. First, the new extended supersymmetric Standard Model incorporates a variety of phenomena that were mysterious before it emerged, and second, the supersymmetric Standard Model is one effective theory closer to the final theory. If nature is indeed supersymmetric, explaining the origin of supersymmetry breaking will become the central problem of supersymmetry theory and experiment. **In the compactified M-theory, the supersymmetry breaking is required by the theory, unavoidably.**

The manner in which supersymmetry explains the Higgs physics is elegant and has important consequences for how we expect to test supersymmetry experimentally. It is rather technical. There are three parts to the Higgs physics of the Standard Model. First the Higgs field must exist. Then it must interact a certain way with the other particles—that is called the Higgs mechanism. Third, the quanta of the Higgs field, one or more Higgs bosons, must exist—that is required once the Higgs field exists. The supersymmetric Standard Model automatically contains fields like Higgs fields, so it makes their existence more natural. Supersymmetry also provides the interaction, the Higgs mechanism. The crucial signal that the Higgs mechanism is working is that a parameter of the theory that naïvely should be interpretable as the square of a mass (let's call it M^2 to give it a name) and therefore a positive number, is in fact negative. While that seems at first to be a difficulty, it turns out on deeper examination to do what is needed to give the other particles a mass. So one can tell if the Higgs mechanism is working in any theory by checking to see if the quantity called M^2 is negative. In the supersymmetric Standard Model M^2 is one of the supersymmetry parameters, and has a positive value at the Planck scale so all the quarks and leptons and W and Z are massless at that scale. But quantities such as M^2 aren't constant—rather, they vary as the theory is applied at different distance scales. We need to know M^2 at the scale of the weak interactions and larger scales, where we live and where our experiments show that the quarks and leptons and W and Z have mass. In the supersymmetric Standard Model we can calculate how the value of M^2 changes as it goes from the Planck scale to the weak scale. The remarkable result is that if one condition holds M^2 decreases to zero and becomes negative at larger distances, inducing the Higgs mechanism!

That condition was that one of the quarks had to be rather heavy, heavier than a W boson. When the way supersymmetry could explain the Higgs mechanism was first pointed out in the early 1980s, the heaviest known quark was far lighter than the W boson, but the mass of the top quark was not known. So supersymmetry predicted that the top quark would be far heavier than the naïve estimates. When the top mass was finally measured in the 1990s it was indeed about twice the W mass, comfortably satisfying the supersymmetry prediction. If history had been a little different and supersymmetry had been developed and understood before the Standard Model, the Higgs mechanism would not seem mysterious at all, but just a new and unexpected consequence of supersymmetry.

The way supersymmetry explains the Higgs mechanism leads to an important result—it is the only way we know to relate the masses of the superpartners to known masses, so we can estimate how large the superpartner masses are. The Higgs mechanism leads to a description of the masses of the Standard Model particles in terms of the masses of the superpartners. So one obtains an equation with a known Standard Model particle mass (W or Z mass) on one side, and unknown superpartner masses on the other. For any equation like that one would not trust the result if the quantities on one side were much larger than those on the other, because any measurable quantity in a physics theory can only be estimated to some accuracy; there are always experimental and approximation errors involved. Therefore, the supersymmetric Standard Model explanation of the Higgs mechanism would not make sense unless the superpartner masses were not much larger than the Standard Model masses they explain. That gives us an estimate of the masses we should expect the superpartners to have as we search for them.

IOP Publishing

String Theory and the Real World (Second Edition)
The visible sector
Gordon Kane

Chapter 5

Compactification

We've already learned in the mid-1980s the key fact that underlies our whole modern theoretical structure: basically, **if we want a theory that has a mathematically consistent quantum theory of gravity it must have nine (or ten) space dimensions. The underlying theory with ten space dimensions is M-theory**, and several of the string theories are special limits of M-theory. **We're interested in the case where six or seven dimensions are curled up in a Planck-scale size region**, a manifold. These theories are good candidates for consistent quantum theories of gravity.

Amazingly, when three of the space dimensions inflate so the theory is compactified, the rest of them can behave like the forces of the Standard Model. The number of extra dimensions has a remarkable physical interpretation.

5.1 It's astonishing that a mathematical physics argument has pushed us to think we live in a world with nine or ten space dimensions

We'll see that once we take the extra dimensions seriously, we find the case for them gets stronger!

Mathematically, the equations that push us to extra dimensions is simply have a factor $(D\text{-}10)$ in a term whose presence would lead to an inconsistency in the theory, where D is the total number of space–time dimensions, and as always there is one time dimension. Then by setting $D = 10$ the flawed term vanishes.

Now, remember that our world is described by solutions of the theory, so physicists had to look for solutions. There were many places to look—the various string theories, and many compactifications of each of them to worlds with three large space dimensions (and one time dimension related to those by special relativity). Each apparently gave a different theory that might describe our world. There did not seem to be any theoretical way to argue that one of them should describe our natural world better than the others, and people got discouraged in the late 1980s. Many theorists turned to study the mathematical structures of the theories. Today most, though not all, string theorists study theories, and do not look for solutions that describe our world.

In the mid-1990s Witten discovered M-theory. **It was quickly recognized from earlier work that M-theory had a solution with three large space dimensions and seven curled up small dimensions, called a spontaneous compactification. The small dimensions of about Planck-scale size formed a manifold, in this case called a G_2 manifold** (for technical reasons we don't need to know). G Papadopoulos and P Townsend and others soon recognized that that solution automatically was a spontaneous compactification and a supersymmetric relativistic quantum field theory, so **supersymmetry did not have to be assumed or input. That made the compactified M-theory very attractive to study as a candidate for describing and explaining our world.** A number of people pursued that path in the late 1990s and 2000s.

When the string theories (as opposed to M-theory) are compactified, the Planck-scale sized manifold is called a Calabi–Yau manifold. For technical and historical reasons a lot was known about Calabi–Yau manifolds, and physicists studied them and quickly added more. Several people have succeeded in compactifying some of the string theories to imply the Standard Model in 4D, via the Calabi–Yau manifolds.

In the years after Witten discovered M-theory, considerable significant work was done on the physics of compactified M-theory. Although much of it is technical, Bobby Acharya (then at Rutgers, now Professor at King's College London and ICTP, Trieste) developed **a simple physical picture. Imagine the seven-dimensional G_2 manifold filled with many three-dimensional submanifolds, called '3-cycles'. 3-cycles might have a number of massless particles on them, and the particles would interact according to certain rules described by symmetries analogous to the ones of the Standard Model,** an SU(3) set describing stronger interactions that could give bound states, and an SU(2) × U(1) that would describe some other interactions. Each 3-cycle might host a world if inflation occurred on it.

Our 3-cycle is called the visible one, not surprisingly. Initially it has unbroken supersymmetry. Two 3-cycles are unlikely to interact via contact since they move in a 7D space, but they will interact gravitationally. Typically, some 3-cycles will have their supersymmetry broken, and can transmit that supersymmetry breaking to our 3-cycle gravitationally. Such effects can be calculated.

We'll call the various 3-cycles 'hidden sectors'. Some of them will have stable particles, e.g. like the electron in our sector may be stable since it may not have any lighter particles to decay into. **Then it will provide some of the dark matter.** Some of the neutrinos may be stable. Probably the visible sector has no stable massive particles besides the electron (which may have no lighter particle to decay into), possibly the proton, and possibly one or more neutrinos. If we look at the LSP just phenomenologically, it is likely to look stable (with nothing to decay into), and may be the dark matter. But when the hidden sectors are included, they supply final states for LSP decay, and generate 'portals' connecting the LSP those final states.

5.2 String theorists study theories, not phenomena

Generic compactified M-theory results—no parameters (!)

We've already seen that M-theory compactified on a G_2 manifold gives a 4D supersymmetric, relativistic quantum field theory. The 3-cycles lead to Yang–Mills gauge theories just like the Standard Model, and massless quarks and leptons are present, just like the Standard Model matter.

In order to calculate predictions we need to know a function called the Kähler potential, which is like the metric for the scalers, another function called the gauge kinetic function, which is like the metric for the fermions, and the superpotential, which is like the Lagrangian. We work in part of the theory called the fluxless sector. The theory has a symmetry called a shift symmetry, which implies that all the terms in the superpotential are non-perturbative with known content and there are no polynomial terms with perturbative unknown coefficients. **The generic Kähler potential [1] and the generic gauge kinetic function [2] have been derived, so we can just use them.**

Amazingly, a number of major results can be obtained from these, without knowing further details. All these results follow from global properties, and do not depend on any detailed properties.

- ✓ The moduli occur in an exponent, so when it is expanded all the moduli are in a potential formed by powers of other moduli, and all moduli are stabilized.
- ✓ A large number of hidden sectors with singularities occur. The singularities lead to gauge groups of all types, U(1)'s, SU(2)'s, ..., E_6, E_8. The larger gauge groups run faster because of their larger charges, and form bound states, which cause supersymmetry breaking in the hidden sectors. In the seven-dimensional space two three-dimensional 3-cycles are very unlikely to collide, so supersymmetry breaking is transmitted to the visible sector gravitationally.
- ✓ **The theory has Yang–Mills gauge groups and forces, and massless modes for quarks and leptons.**
- ✓ **After inflation ends the moduli will oscillate in their potentials, giving a matter-dominated cosmological history.**
- ✓ Gravitinos and all scalars have masses of about 35 TeV. **That implies they will not be seen at LHC in energy or luminosity upgrades.**
- ✓ Gauginos (gluinos, charginos, photino, LSP) don't get mass from the moduli F-term, so their mass is of order a TeV. **Gluinos could be as light as 1.5 TeV, or up to about 5 TeV. Charginos could be as light as about 600 GeV, and the LSP as light as about 450 GeV, or somewhat heavier.**
- ✓ **Radiative electroweak symmetry breaking occurs. The Higgs sector is the decoupling one, with one light Higgs boson whose mass is M_Z + the contribution of one top loop, about 125 GeV.**
- ✓ **The inflaton was the overall volume modulus.**
- ✓ **The LSP is unstable, decaying to hidden sector matter.**
- ✓ **The Universe has a de Sitter vacuum, because of the F-terms of hidden sector matter.**

✓ The soft-breaking terms can be shown to all have the same phase (at tree level), so it can be rotated away. Electron and quark electric dipole moments (EDMs) therefore vanish at the high scale. The CKM phase in the trilinears induces some EDMs from the RGE running to the electroweak scale, which can be calculated.

✓ The predicted EDMs are very suppressed, and were predicted before the recent reported data.

[The superpartner spectrum depends on the parameter called μ, which depends on the internal properties of the moduli.]

It is astonishing that all these results follow from the generic kinetic function and the generic Kähler potential and the superpotential with a shift symmetry, all of which contain no adjustable parameters. The compactified M-theory takes us part-way to the goal of a top-down, parameter-free theory of the world we see.

References

[1] Beasley C and Witten E 2002 *J. High Energy Phys.* JHEP07(2002)046 (arXiv: hep-th/0203061)
    Acharya B S, Denef F and Valandro R 2005 *J. High Energy Phys.* JHEP06(2005)056 (arXiv: hep-th/0502060)
[2] Lukas A and Morris S 2004 Phys. Rev. *D* **69** 066003

IOP Publishing

String Theory and the Real World (Second Edition)
The visible sector
Gordon Kane

Chapter 6

The visible sector

Suppose that we got lucky and our compactified M-theory actually described nature correctly. What would we mean by that? How could we tell? Could we have criteria such that almost everyone would agree, and agree on calling it the final theory? Steven Weinberg has written a book, *Dreams of a Final Theory* [1]. See also the YouTube talk of Hirosi Ogurri, also titled 'Dreams of a Final Theory' [2]. It's important to understand that the desired theory is one theory that lets us understand many phenomena—in particular, the list below in this chapter. Including a solution to the Hierarchy problem is essential. The compactified M-theory is automatically supersymmetric, though the supersymmetry shows up in a somewhat non-standard way. At least one superpartner should be observable at a HE-LHC, and could be observable at a HL-LHC. We don't know what the dark matter is yet, but the theory has several candidates that could supply the full relic density, so it is satisfactory in that regard. It's well known that supersymmetry cannot be broken in the visible sector, so hidden sectors are required. The 'final theory' is the theory of the visible sector. Hidden sectors may be described by a different theory.

Weinberg didn't actually define the final theory. I think we do need a definition, because there are two kinds of questions. One kind are the traditional questions about describing and explaining our world. It includes the questions the Greeks asked, and scientists starting with Copernicus and then Galileo. It has led to the Standard Models of particle physics and cosmology, and a comprehensive explanation of our world and all we can perceive. I would like to think that eventually we could understand all the phenomena in a given vacuum.

The rules are softly broken supersymmetric relativistic quantum theory. The resulting low scale theory automatically has a UV completion, since it started from one. The particles are quarks and leptons, and the forces are the strong force and the electroweak force, which unify at a high scale and then unify with gravity. They emerge from one E_8 singularity, which implies three families. Everything and every process we see is understood, in physics, chemistry, biology, and so on. In a page or so I'll list a set of questions to be specific. Let's call them **'easy questions'**. I would like to think that in a given vacuum physicists could understand all the questions in the list. I have a separate discussion of the cosmological constant below. A few of these questions are somewhat technical.

Before I do that, I want to emphasize that there is still a set of questions besides the easy ones. We can call them the **'hard questions'**. The hard questions include ones like why are the rules based on quantum theory? How does space–time emerge? What is the solution of the black hole information paradox? One question that is sometimes in the hard questions but I put in the easy ones is that the world is describable in mathematical language. If it were not, it would fall apart. The final theory probably does not yet answer any hard questions; whether it does is probably irrelevant to answering the final theory questions.

The visible sector or final theory list. Suppose someone found a compactification of an $11D$ M-theory on a G_2 manifold that was a good candidate for describing and explaining our world. At what stage would we call it a final theory, i.e. a satisfactory description of the visible sector? **It would automatically be UV complete since we started with a UV complete theory**, in this case $11D$ supergravity.

The Yang–Mills theories are relativistic quantum field theories. Probably we would be happy if they explained a relatively small list of important phenomena. Some of these phenomena are computed [3] using a generic Kähler potential, a generic gauge kinetic function, and a superpotential with a shift symmetry. No adjustable parameters are introduced. Some predictions are included [11].

Here is a list that would satisfy most people (don't worry about a few technical things included for completeness). In brackets a source is given from which each topic can be traced.

6.1 The final theory list

- ✓ **Origin of matter (quarks and leptons)—the Standard Model [Compactification, SM].**
- ✓ **Origin of quark and lepton mass Hierarchy? [4]**
- ✓ **Origin of weak CP violation [SM]**
- ✓ **Origin of forces (strong, electroweak, gravity, unified?)?[Compactification, SM]**
- ✓ **Origin of electroweak scale (Higgs mechanism, allows quark and lepton, W and Z masses)? [3, 12]**
- ✓ **Origin of supersymmetry (or its absence) [Compactification]**
- ✓ **Origin of supersymmetry breaking? [3]**
- ✓ **Origin of the Hierarchy? [3]**
- ✓ **Origin of complete moduli stabilization? [3]**
- ✓ **Origin of scales (supersymmetry breaking; electroweak symmetry breaking; μ; gravitino)? [3]**
- ✓ **Origin of superpartner masses, values? [3]**
- ✓ **Origin of strong CP violation? [5]**
- ✓ **Origin of cosmological history? Matter, moduli, or radiation dominated? [6]**

✓ **Origin of inflation? What is the inflaton? [7]**
✓ **Origin of De Sitter vacuum? [3]**
✓ **Origin of matter asymmetry? [8]**
✓ **Origin of flavors, why three flavors in nature? [4]**
✓ **Origin and nature of dark matter? [9, 10]**

A few of these topics are technical. The reader can get the idea without being able to grasp every one. Several are included so workers in the field don't have to wonder about them. I hope readers will think about the list. *The checks say that M-theory compactified on a G_2 manifold answers the question or explains the phenomenon. Since all the topics are addressed, the compactified M-theory is a candidate for the final theory.*

The theory automatically predicts superpartners for all the particles. The full supersymmetry of the theory is necessarily broken at lower energies, by gaugino condensation [3], with predictions for LHC. Currently LHC has entered the energy region to test the predictions, but not covered most of it. Except for dark matter, all the others are present in the compactified M-theory worked out by Acharya and myself and collaborators starting in 2006. A very important point is that the compactified M-theory automatically contains dark matter candidates, and that generically some dark matter candidates do give the relic density. That's very different from the Standard Model or other theories which cannot have dark matter candidates.

In the late 1980s theorists were discouraged because there were several types of string theories, and several ways to compactify, so how could we hope to find the right one via a top down approach. It turned out that was not so hard. For example, in the compactified M-theory, one noticed that the compactification to four large space–time dimensions plus seven curled up dimensions led automatically to a supersymmetric relativistic quantum field theory, so it was natural to study it further. Forces like the Standard Model ones, Yang–Mills ones, and particles like the Standard Model ones (quarks, leptons, gauge bosons like W and Z) all were present in the theory. Further study showed all the moduli could be stabilized, and supersymmetry was broken in a satisfactory way.

Then one by one we worked out the theory's predictions for various important observables, in particular the ones in the list, over a 15-year period. The Higgs physics was the most important since LHC was taking data. PhD students (Piyush Kumar, Lu, Jing Shao, Bob Zheng), postdocs (Konstantine Bobkov, Martin Winkler), and three senior people (Fred Adams, Malcolm Perry and Scott Watson) focused on the issues, and joined Acharya and/or me in working out the predictions, over a period of 15 years.

The Cosmological Constant. The big question is of course the cosmological constant. Do we have some ideas about its role? Actually, possibly we do.

What about the value of the cosmological constant? We want to understand the world we see, and that should eventually include the cosmological constant. We have shown we can understand the other issues on the final theory list. To approach the problem, we have to think about how the Universe can end up into different vacua. Let's assume the vacuum energy density is due to a cosmological constant. When its value is calculated it turns out to be far too large. The cosmological constant experimental value is ~ $(10^{-3} \text{ eV})^4$. Sometimes people say the calculated theory is too large by a factor of about 10^{120} GeV. That's an exaggeration.

First, the cosmological constant has dimensions of (mass or energy)^4 so the relevant scale is the fourth root. Second, the supersymmetric value for the 4th power is about 10^{44} rather than 10^{120} so the 4th root means **we expect trouble at a scale of about 10^{11} GeV, compared to the experimental value. That's not good, but it's a lot better than 10^{120}**.

[The supersymmetric vacuum energy value is about 10^{44} GeV^4 rather than 10^{120} GeV^4 because the latter is M^4_{Planck} while the former is $M^2_{\text{Gravitino}} M^2_{\text{Planck}}$ with some numerical factors.]

For compactified M-theory in reference [3] we found two significant results when we studied how the theory behaved near the minimum of the potential. Let's call the cosmological constant CC. We set the value of the potential to zero at tree level, so we get an expression for CC = 0. That doesn't solve the CC problem, but it puts a non-trivial condition on the solutions. When we impose that condition on the gravitino mass, for generic G_2 manifolds it turns out that the resulting values of the gravitino mass are all in the tens of TeV region. Thus we do not have to independently set V_0 to zero and set the gravitino mass to the TeV scale as has been required in previous work in string theories.

Second, we can estimate numerically [3] for an example with two moduli how the amount of tuning of the CC might affect the values of the M-theory soft-breaking parameters. Subleading terms lead to a value of $V_0 \sim 0.01 \times M^2_{3/2} M^2_{\text{Pl}}$, but we compute the numerical value of gaugino masses and find it shifts by only about 1%. These two results encourage us to expect that not solving the CC problem will not stop us from doing phenomenology. Of course they are not a solution of the CC problem.

Further, it's not true that we have no idea about calculating the CC, or anything else. Papers 'solving' the CC problems appear on the arXiv, probably more than one a week in recent years (Google it). I was going to say once a month, but I looked and over once a week is closer. There are lots of ideas, but the calculations are difficult. Maybe one of the ideas being put forth does work, but we don't know because the necessary calculations have not been completed. (I proposed an approach I like to solving it over 15 years ago, with M Perry and A Zytkow. With Fred Adams and Scott Watson we even showed it led to a different approach to inflation. But then the compactified M-theory came along, and we thought LHC would turn on soon, so work on the cosmological constant got put aside for a while. I still think it's an interesting idea). As we learn more about the theory, better calculations may become possible. PhD theses may be devoted to risky calculations of the vacuum energy. The arguments of Banks and Fischler that the CC is a boundary condition [13] may turn out to be basically correct, in which case the problem is removed. We have to understand vacuum phase transitions and holography better. I think we should not give up on calculating the CC until much more work is done toward calculating and understanding it.

One can imagine at least two outcomes for how we should think of a small CC. One is that it is there, and it does not significantly affect any other observables, such as the superpartner masses, quark masses, inflation, etc. Not calculating it does not prevent us from calculating other observables. And calculating it does not affect results for other observables. The CC problems are probably not the most important problems in physics and cosmology. Perhaps one could call them the most challenging ones. It is premature to give up on solving the cosmological constant problem(s).

Whatever the outcome that emerges, it shows us again that the boundaries of science have changed. It's important to understand that inputting a few general conditions leads to a successful theory in spite of the large number of possible solutions. One compactified M-theory leads to a UV complete 4D theory with solution of the Hierarchy problem, Yang–Mills gauge theories for forces, quarks and leptons in the massless spectrum, generic gauge kinetic function and generic Kähler potential without adjustable parameters to the extent calculations require, and descriptions or explanations for the main observed phenomena.

Uniqueness? If we find a strong candidate for a final theory because it describes and explains the world we perceive, including the facilities to probe the smallest and largest parts of the Universe, should we expect to find some other compactifications that also do that? Or should we expect that our final theory is *the unique final theory?* I prefer the latter alternative for our sector, but I don't have strong arguments for that so far.

References

[1] Weinberg S 1992 *Dreams of a Final Theory* (Pantheon)

[2] Ooguri H 2012 *Dreams of a Final Theory* YouTube, TED talk

[3] Acharya B S, Bobkov K, Kane G L, Kumar P and Shao J 2007 Explaining the electroweak scale and stabilizing moduli in M theory *Phys. Rev.* D **76** 126010 (arXiv: hep-th/0701034)

[4] Gonzalez E, Kane G, Nguyen K D and Perry M J 2002 Quark and lepton mass matrices from localization in M-theory on G2 orbifold arXiv: 2002.11820 [hep-th] (to appear in *Phys. Rev.* D)

[5] Acharya B S, Bobkov K and Kumar P 2010 An M theory solution to the strong CP problem and constraints on the axiverse *J. High Energy Phys.* JHEP11(2010)105 (arXiv: 1004.5138 [hep-th])

[6] Acharya B S, Kumar P, Bobkov K, Kane G and Shao J 2008 Non-thermal dark matter and the moduli problem in string frameworks *J. High Energy Phys.* JHEP06(2008)064 (arXiv: 0804.0863 [hep-th])

[7] Kane G and Winkler M W 2019 Deriving the inflaton in compactified M-theory with a De Sitter vacuum *Phys. Rev.* D **6** 066005

[8] Kane G and Winkler MW 2020 Baryogenesis from a modulus dominated universe *J. High Energy Phys.* JHEP02(2020)019 (arXiv: 1909.04705 [hep-th])

[9] Acharya B S, Ellis S A R, Kane G L, Nelson B D and Perry M 2018 Categorisation and detection of dark matter candidates from string/M-theory hidden sectors *J. High Energy Phys.* JHEP09(2018)130 (arXiv: 1707.04530 [hep-th])

[10] Acharya B S, Ellis S A R, Kane G L, Nelson B D and Perry M J 2016 The lightest visible-sector supersymmetric particle is likely to be unstable *Phys. Rev. Lett.* **117** 181802 (arXiv: 1604.05320 [hep-ph])

[11] Ellis S A R and Kane G L 2016 Theoretical prediction and impact of fundamental electric dipole moments *J. High Energy Phys.* JHEP01(2016)077 (arXiv: 1405.7719 [hep-th])
[12] Kane G, Kumar P, Lu R and Zheng B 2012 Higgs mass prediction for realistic string/M theory vacua *Phys. Rev.* D **85** 075026 (arXiv: 1112.1059 [hep-ph])
[13] Banks T and Fischler W 2018 Why the cosmological constant is a boundary condition arXiv: 1811.00130

IOP Publishing

String Theory and the Real World (Second Edition)
The visible sector
Gordon Kane

Chapter 7

Anthropic questions and string theory

If the laws of nature were different could life nevertheless exist? Does the Universe have to be a certain way for us to be here? More precisely, the particles and forces that determine what happens in the Universe have various properties and strengths and ranges— if those were different what would happen? If Planck's constant or the speed of light were different, would the world be different? These questions were first raised in this modern form in the 1960s and 1970s. It was initially argued that if any aspect of the laws and constants of nature was any different from what we observed it to be, our Universe would be very different and life would not exist. For example, there is an attractive strong force between a neutron and a proton, so that they bind into a deuteron, the second nucleus of the periodic table. From the point of view of the strong force, neutrons and protons essentially behave identically, so there will also be an attractive strong force between two protons. But since the two protons are both electrically charged, they will also feel a repulsive electrical force. In our world that repulsion is sufficient to keep the two protons from binding. However, if the strong force were a little stronger, two protons would bind in spite of the repulsive electrical force (unless the electromagnetic force also got stronger, e.g. as in grand unified theories). Then the reactions that power the Sun would proceed at different rates; the Sun would burn its fuel more quickly, and there would not be time for life, dependent on that energy, to evolve on planets. Such arguments are called 'anthropic'.

Such anthropic questions are clearly not ones which can be answered by experiments, but they are nevertheless research questions that can be addressed from the theory side. If we had the final theory we could work out the answers to such questions. Even assuming only that we will one day have a confirmed string/ M-theory, we can address a number of anthropic questions. They should be addressed because usually the arguments are oversimplified, because many anthropic questions are addressed by having more complete theories, and also because non-scientific and even religious interpretations are increasingly being given to anthropic arguments whose validity are not established. To quote one example,

doi:10.1088/978-0-7503-3583-6ch7

Vaclav Havel, former president of the Czech Republic and a noted writer, has said 'we think the Anthropic Cosmological principle brings us to an idea perhaps as old as humanity itself: that we are not at all just an accidental anomaly, the microscopic caprice of a tiny particle whirling in the endless depths of the Universe. Instead we are mysteriously connected to the entire universe...'

Anthropic arguments can be split into two kinds. One simply takes account of the fact that people exist, so the Universe has to be old enough to allow stars and planets to exist, and people to evolve. This is not controversial scientifically—that people exist is data to include as we try to understand the Universe. It may be that lots of universes exist, and only some have the properties that allow people to exist. We can call this kind of anthropic explanation 'minimal anthropic.' There are other kinds of anthropic arguments and explanations that are sometimes discussed; in order to avoid lengthy comparisons we will not examine those here. For our purposes it is sufficient to split anthropic arguments into 'minimal' (defined in this paragraph) and 'non-minimal' (all the rest). An example of a non-minimal anthropic argument is the claim that human life would not exist if the strength of the strong force had a slightly different value from its actual value, and that why it has the value it does cannot be explained scientifically, implying it has that value so that humans can exist, or even must exist.

One strong reason not to take non-minimal anthropic arguments seriously as implying anything about the world being somehow right for human life is the past existence of the dinosaurs. The earth was a suitable place for them, and they were a dominant species for about 150 million years (nearly three times longer than mammals, and one hundred times longer than humans). But for a chance asteroid impact 65 million years ago, they would still be the dominant species, and mammals (including humans) would not have been able to evolve to their present forms. Any argument about the meaning of the Universe should apply equally to the Universe 100 million years ago and now. If the Universe was planned for humans somebody got it wrong. Indeed, perhaps one day all humans on earth will be killed by an asteroid impact—the probability for that to happen is not negligible. If that did happen, would it change how the physical universe and its origin should be explained? The Universe would go on, not noticing. Nevertheless, let us examine the non-minimal anthropic issues more technically because they raise interesting questions and suggest issues for the final theory to solve.

Can one conclude that somehow the strengths of the forces are set to their values so that human life will exist? Some people have drawn that conclusion. But most scientists believe that such a conclusion is unwarranted by the evidence. First, no one has done the full calculations to determine what actually would happen if the strength of all the forces were different. (But see Fred Adams, Physics Reports [1], described just below.) Often in a complicated physical system new and subtle interplays at different stages can enter to change the outcome of a calculation. The strength of the strong force enters at a number of places in the full set of processes that fuel the Sun. How much can the force change before there is an effect —1%? —10%? For example, when a major feature of an ecosystem changes, the ecosystem might die, but more often it just readjusts and goes on.

Second, and more important, almost all scientists expect that scientific explanations will be found for all or almost all of the non-minimal anthropic questions. For example, if we did not understand evolution by natural selection, we might think that the way the human body and mind work are evidence of design or planning, an anthropic explanation. But now after a century and a half of study, extensive documentation has been found to confirm the evolution of the human eye and brain and body. Non-minimal anthropic explanations should not be accepted, nor should any explanation, until alternatives have been explored and found not to do the job. Actually, what kinds of explanations one accepts is a more subtle issue. Nobody can stop someone who prefers to explain nature by saying it was designed for humans to exist. But such an explanation is not scientific, and therefore there is no reason for anyone else to accept it. The question is whether there are scientific explanations, that can be confirmed by any trained person, that are sufficient, and are consistent with the rest of our scientific description of nature. If all non-minimal anthropic issues have scientific explanations, then non-minimal anthropic arguments cannot be interpreted to imply the Universe is a just-so place for humans.

For example, we have seen that in a unified supersymmetric Standard Model, and in string/M-theory, the forces of nature are found to be unified. That means that their ratios are fixed by the basic structure of the supersymmetry theory. Then if the strength of the strong force is increased, so must the strength of the electromagnetic force be increased, or the theory would not be a consistent one. If you increase both forces, deuterons are bound more tightly, but protons also repel more strongly, so the behavior of stars will change less, perhaps very little. When several quantities vary each will be able to have a larger range. The description of the law(s) of nature in a given string/M vacuum may be unique, though perhaps it could be different in other string/M vacua. If that is so, why forces have the values they do is explained without reference to whether the Universe contains life.

Sometimes proponents of non-minimal anthropic effects try to make the case more compelling by stating it in terms of probabilities. If each force, and several other things, all have some small probability of being in the range needed for human life to occur, and the probabilities are independent, then the probability they would all be in the required range is the product of the probabilities and therefore very small. When that argument was first made three decades ago it was worth examining. But now we understand that the forces are unified, so if one force is in the right range the others must also be. The probabilities are not independent, and should not be multiplied, so they are not as small as has been often claimed.

Another non-minimal anthropic issue is the need for neutrons to be a little heavier than protons, which basically translates into requiring up quarks to be lighter than down quarks; those masses are calculable in string/M-theory, though they have not yet been uniquely calculated. If the world is described by a unique string/M-theory this outcome will not be independent of the others. So even if someone wants to claim that some aspects of the Universe we have not yet understood are non-minimal anthropic ones, their probabilities should not be multiplied if they are correlated in the theory, so the combined probabilities would not be nearly so small as is usually claimed.

New results about 'anthropic calculations'—Fred Adams, Physics Reports

Until we have the final theory and understand how to calculate its implications we cannot settle all anthropic questions, so they are valid and interesting issues to study. In spite of that and the obvious interest of anthropic issues, most physicists do not study them because they expect there will finally be few, if any, anthropic effects that will not be better understood by normal physics. There are, however, two very weak senses in which many more physicists expect anthropic effects to occur. One is simply the observation that our Universe must have properties consistent with the emergence of human life, i.e. be minimal anthropic. That is simply taking the data into account, and can have no implications for how we give meaning to life and the Universe. For example, life will obviously not evolve until stars and planets have formed. In addition, heavy elements are essential for life, and we know that the heavy elements are made in exploding stars (supernovas) or colliding neutron stars. So the Universe has to be old enough for the first generation of stars to form and grow old and die, and a second generation of stars to form with planets, before life can evolve.

The second is more challenging. While we do not yet understand how the Universe actually originated, attractive ideas have emerged that provide mechanisms for universes to arise from nothing at all. This is now an active research area. For example, at Stephen Hawking's 70th birthday celebration, which included a conference, six speakers discussed this topic. Perhaps universes might first occur as tiny Planck-scale size entities, and then inflate in a tiny fraction of a second to sizes large enough to see (were someone there to watch). Their total energy including gravitational attraction is zero, but a large part is in potential energy and that is released in the form of particles—that is what we call the Big Bang. These ideas imply that new universes are being created by such processes randomly all the time. It may be that the laws of nature are different in different universes. And perhaps the constants that enter into the laws such as force strengths, Planck's constant h, the speed of light c, the gravitational force strength G, and the cosmological constant (really it is only dimensionless ratios of such constants that should be considered, but we will not concern ourselves with that) can be different. Then life may emerge in some universes but not in others.

Even if that is so, it does not imply any anthropic uniqueness for our Universe. To understand that, think about a truly random lottery. If there is a big prize and many people enter, the chances that any particular person will win are very small (i.e. the chances that any particular universe will be right for life may be small). However, someone is guaranteed to win (i.e. some universes will have the right properties for life). The person who wins may feel lucky, and may 'thank God,' but from the broader perspective we know there is no reason to impute any meaning to which person held the winning ticket since someone had to win. Similarly, those who end up in one of the universes with life may want to feel some uniqueness, and perhaps even to 'thank God,' but at least from the anthropic point of view that cannot be justified. We don't yet know whether the Universe, and the final theory, and the constants and the strengths of the forces are basically unique. But science is moving toward such understanding. With compactified M-theory we may have gotten to the final theory stage.

Recently, Fred Adams produced a powerful review article [1] in which he carefully examines essentially almost all the arguments for any quantities being fine-tuned. For example, quoting him, 'this work removes the unnecessary restrictions that deuterium must be stable and that diprotons must be unstable ... recent work shows that stars—and hence universes—can operate with either stable diprotons or unstable deuterium'. 'Stars can provide both energy and nucleosynthesis in universes without stable deuterium.' Many earlier works have (wrongly) relied on these arguments. Adams shows that careful analysis implies that the allowed ranges for almost all parameters are *much* larger than previously thought. Basically, the only fine-tuned constraint on the parameters of the Standard Models is that the up quark must be somewhat lighter than the down quark. [If not, protons might be heavier than neutrons and therefore allowed to decay into neutrons, and few protons would be left to form hydrogen atoms in the early universe. Then no stars would form.] As far as I know, no theory so far has implied that the down quark must be heavier than the up quark without assuming that result.

Reference

[1] Adams F C 2019 The degree of fine-tuning in our universe—and others *Phys. Rep.* **807** 1 (arXiv: 1902.03928 [astro-ph])

IOP Publishing

String Theory and the Real World (Second Edition)
The visible sector
Gordon Kane

Chapter 8

The scales we need to explain

Only two energy scales occur naturally in physics (the Planck scale, and zero). We must explain all others that arise! That's clear, because other scales would have units that would have to have some origin. The Planck scales have their origin in the fundamental scales, but there are no other fundamental scales. It is striking that most of the evidence for new physics does not determine the energy or mass scales at which the new physics occurs, not for dark matter, or the matter asymmetry, or neutrino masses, [or the supersymmetry μ parameter].

Once the scale of supersymmetry breaking is known for a compactification, much more can be calculated. The superpartner masses are then predicted, and the moduli masses (see below for a description and explanation of moduli), which may greatly affect the cosmological history.

The quantum theory of gravity of course includes the quantum of the gravitational field, the massless graviton, analogous to the photon for electromagnetism. The superpartner of the graviton is called the gravitino. When supersymmetry is unbroken the gravitino is also massless. When supersymmetry is broken the gravitino becomes massive—the splitting between the graviton (always massless) and the gravitino is a measure of the strength of the supersymmetry breaking, and it sets the scale for all the superpartner masses.

In the compactified M-theory there will typically be many hidden sectors. Most will have associated symmetry groups that are U(1)'s, SU(2)'s, etc. Some will have large gauge groups such as E_6 or SU(6). Those with large gauge groups will have large charges in the renormalization group equations, and run faster to get non-perturbative, and break the supersymmetry, at scales around 10^{11} to 10^{13} GeV.

It is important to understand that there are two measures relevant to understanding supersymmetry breaking, one the scale at which it is technically broken (about 10^{12} GeV as described above), and the other the resulting gravitino mass. In the compactified M-theory case the gravitino mass is calculated, and comes out to be about 35 TeV (35 000 GeV). Sometimes even experts confuse these two scales if they are speculating about supersymmetry breaking without a real theory to calculate.

Thus 35 TeV is the natural scale for superpartner masses. That is not a surprising number in a theory starting with everything at the Planck scale, but it is surprising if one expects the superpartner masses to be near the particle masses (which are all well below 1 TeV). **The squarks and other masses are indeed predicted to be at the gravitino scale, tens of tera-electronvolts.**

The theory has formulas ('supergravity formulas') for all the masses. When one calculates carefully the masses of the superpartners of the gauge bosons that mediate the Standard Model forces they turn out to get no contribution from one large source, and **the resulting value for the superpartners of the gauge bosons (gauginos) is on average about 1 TeV, rather than about 35 TeV. They are the gluino, photino, zino, and wino. The strong force gluino is heavier, about 1.5 TeV or somewhat more, up to about 5 TeV, and the electroweak ones (photino, zino, wino) are somewhat lighter, about half a TeV.**

The lightest superpartner, which is important for how to detect the signals at the LHC and for understanding dark matter, will be a combination of the electroweak ones, and thus about half a TeV in mass. All of these are observable at the LHC if it collects sufficient luminosity. Those runs are supposed to collect an amount of data measured in units called inverse femtobarns (fb^{-1}). At the time of writing (December 2020) it has collected about 40 fb^{-1}, and is into the region where we can hope for signals of gauginos but probably 100 fb^{-1} or more data is needed. The goal for the LHC is to collect 300 fb^{-1}. Without a detailed theory to calculate with, we would not have had serious predictions for masses.

8.1 Higgs physics—electroweak symmetry breaking—the supersymmetry Higgs sector

The Standard Model Higgs physics can be loosely thought of as a single Higgs field and a single Higgs boson. The material in square brackets [ ] is technical and can be skipped. The Standard Model Higgs is actually more complicated, so for readers who might want additional detail I will summarize the argument. The Standard Model Higgs fields have to be an electroweak doublet, like up and down quarks, or electron and neutrino. In addition they have to be complex fields. That gives four real Higgs fields. That is exactly what is wanted, because giving mass to the $W^{\pm}$ and Z bosons means giving them a longitudinal polarization state, and three longitudinal polarization states are needed (for W^+, W^-, and Z), leaving one physical Higgs boson as a new particle, as has been observed.

[In the supersymmetric case the symmetry requires two electroweak doublets of complex fields, so eight total fields. There are still three to allow the $W^{\pm}$ and Z to be massive, so there are five physical Higgs states in the spectrum. The actual Higgs boson that emerges as the quantum of the field that breaks the electroweak symmetry can be any combination of these. The data from the LHC tell us that the observed Higgs boson is indeed the one that breaks the electroweak symmetry. The key point is that the decays h → W^+W^- and ZZ are actually forbidden in the

Standard Model if the electroweak symmetry is unbroken [with two Ws each, a triplet state cannot combine to make a doublet Higgs state]. Yet it is observed at full strength, telling us that the actual vertex in the theory is an hhWW one, with one h obtaining a vacuum value different from zero and leaving an effective hWW vertex. The other Higgs states survive and exist as particles but do not participate in the electroweak breaking much (although remember that the errors on the reported data still leave room for some deviation from this conclusion).]

It is very important to recognize that the Higgs boson decay branching ratios tell us that the Higgs field responsible for electroweak symmetry breaking has indeed been discovered, via its quantum the Higgs boson. Given its interaction strengths it seems to be a fundamental particle.

The actual Higgs boson is light compared to the gravitino because of a long familiar property of the supersymmetric Higgs fields called 'decoupling', first recognized by H Haber and Y Nir in 1990, and studied in detail by Haber and J Gunion in 2003. When the Higgs states are heavy one of the two doublets effectively stays at the high mass, and no longer affects the behavior of the light Higgs system. **The light mass can be calculated, and comes out to be 125 GeV.** This analysis was performed in 2011 by myself and my collaborators, before the LHC data were reported. Actually, what is calculated is the ratio of the Higgs boson mass to the Z mass, not the full Higgs boson mass itself. [Technically what is calculated is the coefficient λ of the Higgs field fourth power term at the high scale. It is run to the low scale using a match and run technique.] One can show that the solutions can satisfy the radiative electroweak symmetry breaking conditions.

Actually, it does not matter that it was done before the data, because the analysis has no freedom. The theory is written at the compactified scale, and the full 4D supersymmetry breaking Lagrangian is calculated there. From that stage on the electroweak-scale effective theory is worked out stage by stage, with no undetermined parameters or freedom, so the answer comes out the same whenever it is done. The light Higgs boson mass is correctly predicted by the compactified M-theory, with heavy squarks. On the right-hand side of the overview of scales it is this analysis that brings the Higgs mass down to below a TeV.

Other scales

Any theory that should be taken seriously has to account for several scales in nature, even though the theory starts with a form that has only the Planck scale, and for some a scale of zero. Where can the scales come from?

Mostly we don't have to think about the numerical values of the scales, but we do need to include those in this section. We can use approximate values. Let's use units of electron volts, the energy an electron would get from being accelerated by a potential difference of one volt, 1 eV. Most flashlight batteries are 1.5 volt. A proton has a mass energy equivalent of 934 eV.

The most dramatic one is the electroweak scale. Perhaps one should think of it as the Higgs field vacuum expectation value, which is 240 GeV. A GeV is 10^9 electron volts. The Planck scale is 1.2×10^{19} GeV. The Higgs potential energy has a minimum value of 240 GeV, away from the origin of the Higgs potential. The Higgs potential is calculated from the Lagrangian of the theory, first at the Planck scale, where it looks quite different, and in particular does not have a non-vanishing value in the vacuum.

Recall that when we work with a relativistic quantum field theory the Lagrangian changes with the energy being probed. One of the amazing properties of compactified M-theory is that as the probing energy decreases it gets to a stage where the Higgs potential energy has a minimum away from the origin, and then moves to be 240 GeV. This is the Higgs mechanism! We do not put the Higgs mechanism into the theory so that particles get mass, and the Higgs mechanism does not operate at the Planck scale, but it changes with the energy probed so that the Higgs potential gives the Higgs field a non-zero value away from the origin, at a probing energy of the electroweak scale. [This is called 'radiative electroweak symmetry breaking' for historical reasons.] The masses of the carriers of the weak force, the W and Z bosons, are fixed by the electroweak scale.

The fact that the interaction strength varies with the probing energy leads to other effects. As the probing energy decreases, for the strong force the interactions of quarks get stronger. When the probing energy gets down to the 1 GeV region neutrons and protons and other hadrons form out of the quarks. That explains why the proton mass has the value it does.

A similar effect is responsible for breaking the supersymmetry. The hidden sectors have coupling strengths associated with the symmetry groups on them. There are many tens of hidden sectors for each G_2 manifold, with SU(2) groups, SU(3), ..., SU(6), E_6, E_8, and so on. When two interactions of large gauge groups occur they can form a bound state. That removes them from the natural distribution, and they often bind. There is a certain probability such binding will occur, and generate the associated supersymmetry breaking. That occurs at a rather high scale, about 10^{13} GeV.

We know that supersymmetry is a broken symmetry since superpartners are not found with the same masses as the particles. Supersymmetry will still be relevant to our physics if the superpartners *exist*, even if their masses are different from the partner masses, which is what we mean by a broken symmetry. For example, the way supersymmetry solves the Hierarchy problem will still work if the mass differences are not too large. Since superpartners have not yet been found at LHC, we know lower limits on their masses, but not upper limits.

How heavy might the superpartners be? We have no definite theory of their masses, since they are physics beyond the Standard Model. **The best we can do is construct models which are UV complete so the effects of gravity and the Planck scale are included, and which have a Higgs mechanism that behaves more or less correctly and gives a correct Higgs mass, and see what superpartner masses emerge.** A few people have done such analyses, and the answer is that **gluinos have masses from about 1.5 TeV up to about 5 TeV.** Sometimes for string theories the models have parameters and can be pushed to make lighter or heavier superpartners, but those values may not be typical, but the range 1.5–5 TeV is typical.

The bottom line is that no one should have expected gluinos to be lighter than about 1.5 TeV in realistic worlds, because the calculated masses in the generic models have that property.

8.2 Gravitino and heavy superpartners

The graviton is the quantum of the gravitational field, and mediates the gravitational force. It's massless, and has spin 2. Its superpartner the gravitino is also massless when supersymmetry is unbroken, and has spin 3/2. It mediates supergravity interactions. When supersymmetry is broken the gravitino becomes massive. The masses of the other superpartners generically will be proportional to the gravitino mass.

People speak carelessly about the scale of supersymmetry breaking. As described above, supersymmetry is broken at a scale of about 10^{12} GeV by gluino condensation, This sets the true scale. When the gravitino gets its mass from the supersymmetry breaking, the gravitino mass can also be thought of as a measure of the amount of supersymmetry breaking. It will have a mass of order 35 TeV.

Naively then, the superpartners would have masses of about 35 TeV. That's true for scalars, and approximately for the trilinear couplings, from the soft-breaking

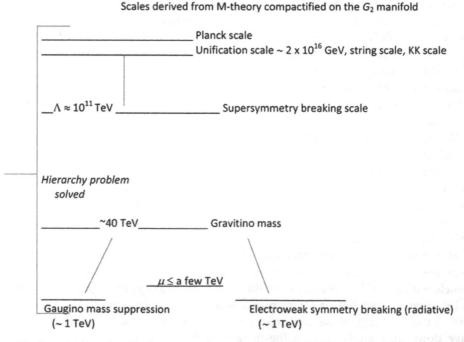

Figure 8.1. An overview of scales that emerge from the compactified M-theory. The different scales are explained in the text. The scales of the quark and lepton masses, and the neutrinos, are not explained here. The natural scale is the gravitino mass, about 35 TeV. The theory tells us that gaugino masses are suppressed compared to the gravitino mass, and the light Higgs boson is suppressed to below a tera-electronvolt. The theory implies this figure for lower scales. This is the answer to the Planck scale being far away.

Lagrangian. It turns out that for gauginos (gluino, chargino, photino, LSP, etc) the leading term vanishes, so gaugino masses are about half a TeV for uncolored ones and about 1.5 or larger for colored ones. The calculations for these are technically challenging so we just report them here. There is another effect that reduces some scalar masses—the running of the couplings down to the TeV scale reduces the third family masses significantly, down to perhaps two thirds of the gravitino mass. Then they have a larger production rate. We'll come back to focus on the light gauginos later.

Figure 8.1 shows some scales, and how they emerge from the Planck scale. [It shows another scale, called 'μ'. μ is important, and related to the Hierarchy problem too, but requires a technical treatment, and we will ignore it here.]

Other scales are associated with the quark and lepton masses, which are very small on the scale of figure 8.1. They arise from different mechanisms, rather technical ones. I'll briefly describe them in a later chapter.

IOP Publishing

String Theory and the Real World (Second Edition)
The visible sector
Gordon Kane

Chapter 9

Testing theories in physics, including string theories

Much has been written and said about testing theories, and in particular much about testing string theories. We'll presume that an essential ingredient of a final theory is likely to be string/M-theory, so we need to focus on testing string/M-theories. We'll explore in additional detail what string/M-theories are as we describe ingredients of a final theory below. Much of what has been said and written on these topics is rather shallow or misleading and not very interesting.

For example, string theories are naturally formulated at an energy scale of about 10^{16} TeV, near the Planck energy scale. That's a very high energy. The CERN Large Hadron Collider, the most energetic facility in the world today, reaches an energy scale of only a little over 10 TeV, and discussions are underway at CERN and in China to eventually build a collider with close to 100 TeV energy collisions. It will not be possible ever to have collisions with energies that could create new particles and test collisions near the Planck scale. Even if there were occasionally isolated particles in the Universe (cosmic rays) of such high energies, their collisions would not be seen in a detector, and there would be far too few to learn anything relevant. Is that a problem for testing string theories? A number of people have said so.

No. Absolutely not. Of course if such facilities could be built they would provide good tests. But even without them, there are always relics. Consider the Big Bang. There were no people to observe it. But all knowledgeable scientists agree it happened, because there are several major calculable relics that provide tests. One is the expanding Universe itself. A second is called nucleosynthesis. Using Standard Model physics and a universe having a hot Big Bang physicists can work out accurate values for the abundance of hydrogen, helium, and several other elements. The predictions agree with the observations with an accuracy of a few per cent. We can see that stars throughout the Universe are made of the same chemical elements we are, because of their emission and absorption spectra.

A third major Big Bang test is the cosmic microwave background radiation. By today the Universe has cooled to a low, predictable temperature (2.73 K). Looking at the sky is like seeing the embers of a dying fire. The Universe looks the same in every direction, as it should if it came from a hot tiny space at the beginning. The temperature everywhere is the same to about a part in 100 000, and it is indeed that predicted temperature. There are additional tests of the Big Bang. All the tests are confirmed with data. The crucial thing to understand is that scientists can always work out indirect tests and implications of any idea, and eventually test it.

Another example is the dinosaur extinction. No one thought we would ever know what happened to them 65 million years ago. Humanoids have populated the Earth since only about a million or so years ago; but amazingly an understanding of nuclear physics and of near-earth astronomy eventually led to learning how an asteroid hit the Earth in the Gulf of Mexico and led to that result. We can learn about it even though (almost) no one thinks people were there to observe it. You don't have to be somewhere to test a theory there.

Yet another is that you do not have to travel at the speed of light to learn it has a maximum speed. It follows from experiments and conditions. Even if underlying theories are formulated at or near the Planck scales they will be and are testable.

Is the absence so far of direct evidence for superpartners evidence against string theory, as some people have claimed? The first question to ask them is if super-partners are discovered tomorrow does that establish string theory? One should know that arguments about discovering superpartners are very dependent on how the superpartners decay. Some decay patterns would have allowed gluinos (the superpartners of gluons) lighter than about 1.5 TeV to be detected if they exist, while other decay patterns would have allowed gluinos as light as 1.1 TeV to not yet be detected. LHC experimenters and commentators have generally emphasized the larger exclusions. Well-motivated theories often contain gluinos that would be not yet detected with masses somewhat above 1 TeV, for understandable theoretical reasons. Gluinos in that mass range will be detectable at the LHC in the next few years if it functions as expected.

More generally, where did the predictions for superpartner masses come from? There were no theories predicting the values of superpartner masses. If super-symmetry is to do all the good things it does, it cannot be too heavy. That argument is called 'naturalness'. The opposite of naturalness is having a theory. The natural-ness argument didn't work well. That doesn't imply superpartners don't exist, because theoretical arguments tend to predict heavier superpartners.

For example, the compactified M-theory case we'll examine below predicts that gluinos will have masses of about 1.5 TeV or more, and decay patterns implying about 500 gluino production events will have to be produced for each detector before a signal from them can be seen above the backgrounds that can resemble signal events. (For the Higgs boson discovered in 2012, over 200 000 Higgs bosons had to be produced in each of the two detectors at LHC before a signal was visible above its background.) The compactified M-theory predicts that three kinds of superpartners will be observable if sufficient numbers of collider events are produced at the LHC with its current energy and intensity (gluinos and positive and neutral winos), but that all other superpartners require higher energy or intensity colliders. More importantly, the prediction is that none of these should have been seen in the LHC data so far. Claims they should have been seen would be valid given naturalness arguments, but are wrong in actual theories. Some of us think that is a misuse of the idea of naturalness, but it is the fashionable use.

Let's look more carefully at the meaning of testing theories in physics. Consider the most well-known theory of motion, Newton's familiar second law relating a force F on a particle of mass m and the resulting acceleration, **$F = ma$. Is it testable as a general statement? Absolutely not.** It can be tested for any particular force, such as gravity or electromagnetism, and for a particular mass or set of masses, by calculating the resulting acceleration and comparing with measurement. But it has no general tests. The situation is similar in quantum theory, where the Schrödinger equation (basically $i\partial\psi/\partial t = H\psi$) replaces $F = ma$, and can be tested by solving it for a given Hamiltonian H but not generally. (In relativistic quantum field theories one specifies a function called the Lagrangian rather than the Hamiltonian.)

The situation is string/M-theory is actually analogous. In order to be possible quantum theories of gravity, string theories must be formulated in ten space–time dimensions. That's the discovery that started modern string theory in 1985. What is meant when someone says the goal is to test string theory? One meaning is to test it as a theory about our world; except for string theorists and philosophers, that is the usual meaning. To test it as a theory about our world *obviously* the ten-dimensional string theory *must* be projected onto a four-dimensional world ('compactification'). Further, theories generically have many solutions, so our world should be one of the solutions, and before there is a test such a solution has to be found. Again, it is the same with $F = ma$ and normal physics.

The phrase 'testing a theory' normally has no meaning in physics—only solutions are testable. Systems like atoms will fall into their ground state, and so will the Universe. If someone starts talking about testing a theory, ask them what they mean. By the solution one usually means the ground state of the theory in particle physics, often called the vacuum state. Most people, and even most physicists, do not think much explicitly about distinguishing between theories and solutions that describe the actual physical systems, but care is worthwhile to avoid confusion. In the case of M-theory the underlying theory must be formulated in eleven dimensions, and compactified to four dimensions on a seven-dimensional manifold of G_2 holomony (to say it technically).

Quantum field theories do have a few tests that do not depend on specifying the force or Lagrangian, such as superposition, or the prediction that all electrons are identical to each other because they are all quanta of the electron field (and the analogous result for all other particles). Could there be similar tests for string theory or M-theory before compactification? One might be a calculation called the black hole entropy calculation, but I won't discuss that since it isn't related to experimental observations. Otherwise, none are known. As we will discuss below, an exciting aspect of compactified string/M-theories is that they determine the forces that shape our world, and the particles. Consequently it seems unlikely that general tests of the ten/eleven-dimensional theory will exist, tests that are independent of the compactification, the forces, and the particles.

It should not be surprising that tests depend on assumptions. All tests of physics theories have always depended on assumptions. Galileo had to use inclined planes to slow falling bodies so the time to fall could be measured. Otherwise it was too rapid. Then he could check he was not biasing the result by varying the inclined plane angle. He assumed air resistance could be neglected in order to obtain a general theory of motion. To test string/M-theory we assume a particular corner of the theory, such as M-theory or heterotic string theory, and we assume a particular compactification manifold. In the approach that I and my collaborators use we also assume that the three functions needed to make contact with the real world, a 'gauge kinetic function', 'Kähler potential', and 'superpotential', will give generic predictions, consistent with the results of those who originally derived them. It's important to understand that the tests of the compactified theory do depend on the full ten or eleven-dimensional theory in many ways. Information about the curled up dimensions is not lost. It determines the forces, the particles and their masses, the symmetries of the theory, the superpartner masses, electric dipole moments, many relations between observables, and much more.

If we want to test a theory of our world we have to find it and calculate its predictions. This is another issue where much that is said is confused. People talk of very large numbers of solutions of string theories, and claim it will be hard to find one that could describe our world among the huge number. In fact, *compactified* **theories generically have many realistic features whose presence limits the number of possible theories. These features include gravity, Yang–Mills forces like the Standard Model ones, chiral quarks and leptons that give parity violation, softly broken supersymmetry, decoupling Higgs physics with a light Higgs boson, families, hierarchical fermion masses, neutrinos with Majorana masses, inflaton candidates, a solution of the Hierarchy problem, a solution of the strong CP problem, and more.** Solutions with such properties are easy to find. *It cannot be emphasized strongly enough that generic string/M-theories compactified to four space–time dimensions behave like theories explaining our world.* There is still work to do to calculate sharp testable predictions and compare compactifications, but **a lot of work is already done by compactification.**

The goal is to find a comprehensive theory and connect it to experimental phenomena. Once that is achieved, maybe we can work backwards to understand how inevitable it is.

IOP Publishing

String Theory and the Real World (Second Edition)
The visible sector
Gordon Kane

Chapter 10

Dark matter candidates

In recent decades we've learned that about a quarter of the Universe is composed of matter. The quarter that is matter is about 1/6 matter like us, basically atoms of the 92 natural chemical elements. The rest is called 'dark matter'. Dark matter clumps to form galaxies.

I guess most people now are familiar with the idea of dark matter, so I won't go over the reasons to accept the existence of the dark matter. There are about six strong reasons to accept the existence of the dark matter; I presume the presence of the dark matter is a fact. The evidence is well summarized by Wikipedia.

To give one reason, figure 10.1 (from Wikipedia) shows the velocity of a typical star in a spiral galaxy. Curve A is the prediction if the matter like us makes up the Galaxy, and curve B is both the data and also the prediction if the Galaxy is about 5/6 dark matter, spread uniformly.

We expect the dark matter is made of particles. As of today, winter 2020, we don't know what composes the dark matter.

There are two dark matter candidates that are known to exist in a string/M-theory world. That is, two particle types that are known to exist in a compactified M-theory world, and would behave like dark matter. Then the question is what fraction of the dark matter each contributes. Most people won't have heard of either. Most physicists have heard of one of them, but couldn't tell you what it is.

One is a kind of particle called an axion. String theories and M-theory imply the existence of axions, but the axion masses and interaction strengths have model dependence. There are also other independent arguments that suggest axions should exist, with some favored masses and interaction strengths. The axion dark matter relic density will be proportional to the product of the axion mass and the number of axions.

If an axion is discovered with a certain known mass and interaction strength, it has a separate motivation—it is just right to solve the strong CP problem, the unexpected absence of CP-violating interactions with gluons. Historically this is how

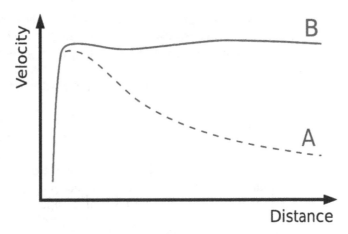

Figure 10.1. The interpretation of the figure and the curves is given in the text.

it originated. CP invariance means that if a process can occur, then the process with charges opposite and also reflected in a mirror can occur with the same rate. Experimentally, gluon processes do have CP invariance, but the theory doesn't require them to—it's surprising that they have that invariance. The interactions with axions is arranged to cancel the relevant terms so they end up with CP invariance. It turned out that the axions could also give the dark matter relic density even though they were not invented for that; an attractive feature.

The second candidate is hidden sector stable matter. As we saw there are lots of sectors. Imagine we are the visible sector. From various symmetries we have some stable matter, electrons, positrons, maybe protons, some neutrinos, etc. Another hidden sector might have different stable particles. **The total dark matter relic density is the sum of relic densities from all the sectors.** If we are lucky, most of the dark matter will be one kind of particle so we can detect it and calculate that we have found all or nearly all of it. But it could easily be the case that the dark matter is the sum of several significant hidden sector contributions.

In the 1980s it was realized that the dark matter could not just be ordinary matter that could not shine, like planets. Initially there was an apparently strong candidate for most of the dark matter, the lightest superpartner (LSP), which could be a photino, a wino or bino, a sneutrino, even a gravitino. Collectively these candidates were called WIMPs, for Weakly Interacting Massive Particles. Note they all had weak or electromagnetic or gravitational interactions with ordinary matter. Innovative detectors were built. As always, initially small detectors were built, then larger mature ones. No confirmed signal was detected even as larger and larger regions of particle type and mass were surveyed with very good detectors. It seemed reasonable for the LSP to be stable, and a good candidate for the dark matter. This was reinforced because calculations with models having a stable LSP usually gave a realistic relic density.

On the other hand, if one imagined a compactified string/M-theory the LSP was almost never stable! One could generalize to say that we are looking for UV complete

theories, so we should always include gravity in our thinking. When we do include gravity, sometimes we get a different result from when we don't include gravity, not just a little different but qualitatively different, even if gravity doesn't enter the calculations technically.

Interestingly, this result emerged independently from the top-down theory and the improving bottom-up data at about the same time.

[What happens in the top-down approach is generic [1]. When the hidden sectors are included, almost always there is a hidden sector particle lighter than the LSP, so the decays are energetically allowed. Again almost always there is a quantum correction loop Feynman diagram connecting the LSP to the final state, so the decay is allowed by a mechanism. The decay lifetimes are typically shorter than the lifetime of the Universe, although sometimes long enough so the candidate LSP can escape the detector.]

Detecting and studying dark matter

For an initial approximation, one can think of the dark matter as a uniform background through the Universe. There are about twenty detectors constructed or being constructed at various locations in the USA, Canada, Europe, and Asia. The detectors are sensitive to various imagined possibilities for the dark matter. The earth is moving in the solar system, and the solar system is moving in the Galaxy. Assume the dark matter is indeed composed of particles. The dark matter particles will have very weak interactions with the quarks and electrons in the detectors. Occasionally a dark matter particle will interact with a particle in the detector, which will recoil. Experimenters have learned how to detect such recoiling particles. If there were no dark matter such signals would not be seen. The first goal will be to see some signal of the dark matter, in one of the detectors.

LHC will also be a factory for the dark matter particles. Think of producing some superpartner or other new object by colliding the protons. The new particle(s) will decay. Eventually there will be a decay to the LSP, which will itself then decay to hidden sector matter. The LHC may be the best way to find hidden sector dark matter. We don't know enough to say for sure that the LSP will not live long enough to escape the detector, though often it should not. Probably the resulting hidden sector particles will escape the detector most of the time, but not for sure, so if the original particle were produced in large quantities it might be possible to study even the hidden sector particles.

Then by varying the materials in the laboratory detectors it will be possible to eventually learn about the properties of the dark matter. It will have different cross sections that determine the strengths of interactions with nuclei in the detector. Currently there are many speculative predictions for dark matter. Once just a little data is known, it is likely the dark matter can be tied to a theory.

Reference

[1] Acharya B, Ellis S A R, Kane G L, Nelson B D and Perry M J 2016 Lightest visible-sector supersymmetric particle is likely to be unstable *Phys. Rev. Lett.* **117** 181802 (arXiv: 1604.05320)

Part II

Explaining and interpreting recent compactified M–theory results

IOP Publishing

String Theory and the Real World (Second Edition)
The visible sector
Gordon Kane

Chapter 11

Moduli

Without learning about compactified string/M-theories we are unlikely to have learned about hidden sectors, or about moduli, both of which play a large role in the phenomenology and cosmology of string/M-theory.

It takes a lot of information to specify the geometry of the curled up Planck size 7D manifold. One has to give the diameters of all seven dimensions, and the angles each plane makes with all the other planes. The values the moduli fields take in the vacuum will determine the observables such as Standard Model particle masses and interaction strengths. An important issue is that the moduli have to take on stable values early in the evolution of the Universe, or the laws of nature would seem to change with time. This is called **'stabilizing' the moduli. The M-theory moduli can all be stabilized.** This is one of the important initial successes that encouraged us to calculate more observables in the M-theory vacuum.

One can imagine inflation ending, with all the moduli interacting and forming a potential for the rest of them. They will fall to the minima of the potential energy and stabilize. Typically their vacuum values will be one to two orders of magnitude below the Planck scale at the minima. The moduli will oscillate around the minima and lose energy by radiating gravitinos and other particles, and decaying. Their masses will be of order the gravitino mass, which will turn out to be of order 35 TeV. The moduli oscillations after inflation will dominate the energy density of the Universe, rather than having it radiation-dominated, which is a major and important difference from the usual radiation-dominated cosmological history. As the Universe size (diameter a) grows, the energy density of matter decreases as $1/a^3$, while that of radiation decreases as $1/a^4$ because the radiation has to fit its wavelength in the growing Universe.

Matter dominated cosmological history

In compactified M-theory, and string theories, there are many moduli fields with masses of order the gravitino mass and larger. The modulus decay rate is

11-1

proportional to its mass (cubed), so heavier ones decay earlier. **When the moduli decay they can decay into everything**—the dark matter, superpartners, Higgs bosons, axions, hidden sector matter etc. The moduli have masses greater than or of order 35 TeV, some up to hundreds of TeV.

At the end of inflation, the moduli find themselves in a potential, oscillating, with some 'friction'. They lose energy as they settle into their minima.

Unfortunately, there is some left-over dark matter from the heavier moduli decays as well as the lightest one, so one cannot just write down quickly one of the most interesting quantities, **the ratio of how much of the matter is baryonic compared to how much is dark.** Experimentally it is about a fifth. That they are so close suggests they have a similar origin, which they do have (moduli decay), but it is hard to calculate. **Most of it comes from the decay of the lightest modulus.**

These two kinds of cosmological histories do lead to different kinds of galaxy formation in the early Universe, and other tests, but there has not been enough study of the predictions to draw conclusions about the Big Bang content so far.

IOP Publishing

String Theory and the Real World (Second Edition)
The visible sector
Gordon Kane

Chapter 12

Hidden sectors

When a string/M-theory is compactified, generically a new and unexpected feature emerges, hidden sectors. **Our world is not the only world, there are generically lots of others. Our world is called the visible sector**. The clue is that it turns out that a careful analysis, familiar to those who have had a course on supersymmetry, shows that supersymmetry cannot be broken in the visible sector! Supersymmetry can be broken in one of the other sectors, and then the breaking is communicated to the visible sector by an interaction between the two sectors.

The hidden sectors take various forms in the different corners of string theory and M-theory. In M-theory Acharya has shown one can think of the hidden sectors as three-dimensional submanifolds floating around in the seven-dimensional main compactified manifold, and those 'three-cycles' have the matter of gauge theories on them—they are visible sectors of other worlds. Our visible sector has the symmetries of the Standard Model on them. Other (hidden)-sectors may have different symmetries, perhaps none, or a U(1) symmetry, or SU(2)×U(1), or SU(2)×SU(2), or SU(3), or SU(6), or E(6), or E(8), any of the Lie groups, with some pattern of occurrences. **In the compactified M-theory there are of order several tens or more hidden sectors, so most likely some of the larger groups occur.**

As we have seen, G_2 manifolds arise naturally in compactifying M-theory, which greatly increases the interest of mathematicians in studying them. You might expect a similar interest in hidden sectors among mathematicians, but in practice there is essentially no interest. For non-singular G_2 manifolds the book of Dominic Joyce [1] contains a lot of information. For singular manifolds we would like to know particularly the distribution of the number of hidden sectors, and the typical gauge groups per hidden sector.

Typical UV completions of the visible sector should include many disconnected pure gauge sectors, with stable glueballs or glueballinos. The decay of the lightest modulus will occur shortly before nucleosynthesis, with a reheating temperature of a few degrees. The compactified M-theory will give a period of late time matter

doi:10.1088/978-0-7503-3583-6ch12

domination followed by a low final reheating temperature from a gravitationally coupled modulus, different from a high reheating temperature plasma, with different phenomenological implications. The **temperature is low enough so baryogenesis cannot occur via leptogenesis or at the electroweak phase transition.**

If your goals include understanding our world, the hidden sectors can't be ignored. They affect the physics outcomes in several major ways, including:

We have already seen that how the supersymmetry is broken is necessarily determined in a hidden sector.

In addition, how the breaking is communicated from the hidden sector to the visible sector can occur in several different ways. Communication by gravitational interactions is always possible, and is expected to be the main way for actual physics.

Hidden sectors can have stable particles just as our visible sector does. Then such particles contribute to the dark matter, and could give the entire relic density. Pure gauge hidden sectors can give glueball or glueballino (gluino–gluon bound states) stable particles.

[There is a visible sector symmetry called 'R-parity' that can have a major role in determining what constitutes the dark matter. If R-parity is conserved, the lightest visible sector superpartner is a strong candidate for being the dark matter. Generically R-parity is broken in the hidden sectors, with the breaking transmitted to the visible sector. Then the lightest visible superpartner will not be the dark matter.]

[If R-parity is conserved the lightest visible sector superpartner could be stable and might be a good candidate for the dark matter. But there will still be a high probability that the lightest visible sector superpartner (LSP) can decay into hidden sector particles and not contribute to the dark matter.]

[Hidden sectors may also explain the matter asymmetry of the Universe. See chapter 14]

It's common to define hidden sectors as any that don't carry Standard Model charges and therefore cannot interact with the visible sector directly, without concern about whether the hidden sector in question is actually a hidden sector of a compactified string/M-theory. It's not known how to do such a check in practice. Studies suggest most phenomenological models are in the swampland.

Reference

[1] Joyce D 2000 *Compact Manifolds with Special Holonomy* (Oxford: Oxford University Press)

IOP Publishing

String Theory and the Real World (Second Edition)
The visible sector
Gordon Kane

Chapter 13

Inflation

It's thought that our Universe begins with an unstable inflating energy density. That's consistent with all present data. Quantum fluctuations that grow with a constant doubling time, of order sixty or so times, will generate a universe the size of order that of a soccer ball. The field energy during inflation is a form of energy inherent to space itself. The energy density stays constant as new space is created by the expanding universe. This form of energy is called 'dark energy'. Cosmologists think of a scalar field, the 'inflaton', doing the inflating. They do not have a physical field with a separate motivation in mind to be the inflaton. To be considered a candidate for a final theory, M-theory must identify what the inflaton is in terms of the fields of the supersymmetric Standard Model, and explain why.

With many moduli and associated fields, M-theory has many scalar candidates. With many candidates, identifying the inflaton seems at first daunting. One can calculate the potential energy of any candidate and ask if it underwent inflation. For almost any candidate, the answer is no. **With a little experience one realizes that the overall volume modulus is the likely inflaton, which turns out to be true**. There are also dozens or more ad hoc models people have proposed for potentials that give inflation, but do not appear in M-theory—they are in the swampland if compactified M-theory is a candidate for the final theory.

There is an inflaton that works in M-theory, a potential with a saddle point giving the slow roll region shown in figure 13.1. Imagine a plot of the potential energy vs the inflaton field $\rho - \rho_0$, where the inflaton field is essentially the linear combination making up the overall volume effective modulus. Imagine that the overall volume effective modulus has fluctuated up to the right-hand corner, and begins to roll down. Around $\rho - \rho_0 = 15$ there is a saddle point, and the inflaton field is quite flat.

[This kind of inflation model is called inflection point inflaton. If one defines an angle α between the major axis of the full volume modulus and the inflaton, the potential of figure 13.1 has $\alpha^2 = 0.76$. This inflaton has over 100 e-foldings. Its observational parameters are $n_s = 0.97$, $r = 5 \times 10^{-7}$. This is a minimal pedagogical

doi:10.1088/978-0-7503-3583-6ch13

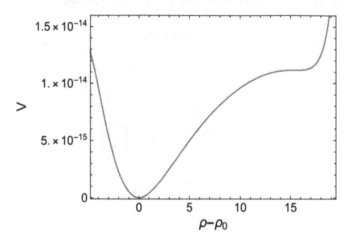

Figure 13.1. Scalar potential in the inflation direction with the other fields eliminated through their minimization condition.

model, since the inflaton must be approximately aligned with the volume modulus, an orthogonal light modulus participates in supersymmetry breaking, and the third modulus is needed for heavy superpartners.]

The view of the Universe has changed greatly over the past century. Just over a century ago leading scientists thought the Universe was unchanging, permanent. Then data from emission and absorption spectra of atoms, and the recognition that atoms are the same everywhere, led to the unavoidable conclusion that the Universe was expanding. Extrapolating backwards implies the Universe had a beginning, just the opposite of unchanging. There are consistency checks—there should be fewer and fewer heavy elements as we look backward in time since heavier elements were formed in exploding and colliding stars; more distant galaxies should be smaller; looking back, there should come a time when the Universe was much hotter, too hot to allow neutral atoms to form; we can deduce when the Universe began; and so on.

Lots of people have talked about inflation. **What is exciting here is that inflation is derived without assuming it. And it is derived in the same compactified M-theory that has a lot of other desirable properties and results.**

IOP Publishing

String Theory and the Real World (Second Edition)
The visible sector
Gordon Kane

Chapter 14

The matter asymmetry

Every particle has an antiparticle, defined to have the opposite value for every charge the particle has, for example, electric charge and any other kind of charge. At the moment when the Big Bang occurred, the unstable energy density turned into matter and kinetic energy. The electron has an antiparticle with opposite electric charge, the 'positron'. Initially there had to be an equal amount of electrons and positrons since the energy density carried no net charge of any kind. Today the ratio of antimatter to matter is extremely small, about one positron for every 10^{10} electrons. The Universe began with an unstable state of energy, so no net value existed of any quantity—no net electric charge, no net rotation, and so on.

How did the huge asymmetry emerge? People figured out in the 1960s that in order to have a theory that could lead to that result it would need four properties. One was that the Universe must not be in thermal equilibrium, so compensation between processes increasing and decreasing the baryon number could not occur. That was satisfied because the Universe was expanding. The theory had to break the matter symmetry to produce an excess of matter. The theory also had to not have a charge-conjugation symmetry so that interactions which produce more matter than antimatter will not be balanced by interactions that produce more antimatter than matter. There is a fourth more subtle condition, that the product of charge-conjugation symmetry and parity symmetry be a broken symmetry, so equal numbers of left-handed matter and right-handed antimatter were not produced, nor equal numbers of left-handed antimatter and right-handed matter. The last condition is separate from the charge-conjugation violation since the weak interactions have left-handed quarks in electroweak doublets and right-handed quarks as electroweak singlets.

Once the grand unified theories were written in the mid 1970s it turned out to be possible to write theories with the needed properties, and actually rather easy. The problem today is that it's too easy, and there are a number of possible models. One can split them into categories according to when they occur as the Universe cools.

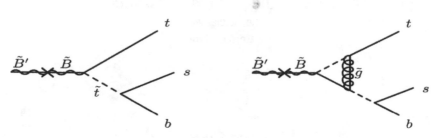

Figure 14.1. The dominant Feynman diagrams will be baryon number violating hidden sector bino decays. The interference between the tree-level and one loop diagrams generates a baryon asymmetry. To get the needed size the phases in the gaugino mass matrices can be a few degrees, and the R-parity violating coupling of order 0.1.

One is at the scale where the couplings unify, just above 10^{16} GeV, right after inflation ends. A second is at the scale of the heavy right-handed neutrinos, about 10^{14} GeV (leptogenesis). A third is at the scale where the Higgs field gets a vacuum expectation value (electroweak genesis), and a fourth at about 10 Mev, just before nucleosynthesis. **In the compactified M-theory we have shown that the asymmetry must be generated at the lowest of these scales, shortly before nucleosynthesis, by a mechanism I'll describe next [1]. In practice, the three higher temperature mechanisms fail in the M-theory universe, and the fourth works.**

First the modulus decays to gauginos with a branching ratio of order unity. Then the baryons are created in the gaugino decay. The needed CP violation originates from the phase of the gaugino mass matrix, which has been generated by the running of the trilinear couplings. Baryon number violation can occur via the R-parity violating three quark operator UDD in the superpotential. Earlier treatments of baryogenesis by gaugino decay were done in a thermal universe. **The thermal universe has difficulty because washout processes (reverse reactions) tend to dilute any generated baryon asymmetry. The washout problem is absent for compactified M-theory because the decay temperature just before nucleosynthesis is far too cold (~10 MeV) for washout processes to matter.** The final matter asymmetry is essentially determined by $\sqrt{M_{\text{MODULUS}} M_{\text{PLANCK}}}$ times one loop integral, where the lightest modulus is relevant. To explain the observed baryon asymmetry of the Universe, takes about 10^{-4} baryons per modulus decay (as shown in figure 14.1).

Reference

[1] Kane G and Winkler M 2020 Baryogenesis from a modulus dominated universe *J. High Energy Phys.* JCAP02(2020)019 (arXiv: 1909.04705)

IOP Publishing

String Theory and the Real World (Second Edition)
The visible sector
Gordon Kane

Chapter 15

Possible tests soon

There are two predictions that compactified M-theory makes that might be tested soon. As is usually the case, tests are seldom simple and definitive. That will probably be the situation here also. Nevertheless, they are instructive and interesting so I will briefly describe them. Both of them may be reported before the book is out or soon after, though COVID-19 delays could complicate that.

The interaction of photons with charged particles can be separated into several pieces that can be studied separately in clever experiments. What makes them interesting is that they can be calculated very accurately in the Standard Model, and also measured very accurately. One of them has the name 'g-2' for historical reasons. It can be studied for both electrons and muons. The other is called an 'electric dipole moment'—it can only be studied well for electrons at this stage.

(g-2)

The interaction of the photon with the muon can spend some time in a virtual state. One of the virtual particles is heavy, and a heavy virtual particle is unlikely to occur compared to lighter virtual states, so the muon, called '(g-2)' for historical reasons, is likely to be very close to its 'tree level' value. [Diagrams with no internal structure are called tree diagrams.] So the compactified M-theory prediction for the muon (g-2) is essentially the same as the Standard Model value.

But the muon (g-2) was measured at Brookhaven National Laboratory in 1992, and found to be about 3.7 standard deviations larger than the Standard Model value. In an ironic reversal, the muon (g-2), which is the only statistically significant deviation from the Standard Model experimentally, is predicted by Beyond-the-Standard-Model physics not to deviate from the Standard Model. What's measured is basically a stronger magnetic dipole moment strength, like the behavior of ordinary magnets.

More recently the electron (g-2) has been reported experimentally to deviate from the Standard Model prediction by about 1.9 standard deviations but to be less than the Standard Model prediction. For the compactified M-theory the electron (*g*-2) one also expects no significant Standard Model deviation.

The Brookhaven experiment was a major effort. A muon beam is put into a storage ring. The muon–photon interaction is measured by keeping track of the muon spin polarization. A group proposed moving the ring to Fermilab and improving the experiment by a factor of four. If the deviation from the Standard Model were real they could confirm it with smaller errors. The storage ring is large, about 30 feet in diameter. The experiment was approved.

The storage ring was uploaded with a helicopter, transported over roads to the ocean near Brookhaven, taken around the waterway and up the Mississippi, and installed at Fermilab! Wikipedia has some remarkable photos.

In principle this will provide a definitive test of the minimal supersymmetric Standard Model, with its prediction that the Fermilab data should show no effect. Unfortunately, however, **in the past year over 20 papers have appeared proposing models with some ad hoc new physics contributions that led to results in agreement with both the Fermilab data and the MSSM theory.** This illustrates how physics works—clear situations are rare initially. Each of the new physics models will be tested, and eventually a clear result will emerge. At this stage we cannot say for sure whether the g_μ−2 data will be important, but we should keep it in mind since it may provide relevant information before new collider results about superpartners.

Electric dipole moments

There is a second laboratory experiment that may provide important information. It involves electric dipole moments (EDMs), which are predicted to be essentially zero for particles and even nuclei. Virtual particle effects can induce EDMs. For the electron, compactified M-theory predicts [1] an electric dipole moment $d_e \leqslant 5 \times 10^{-30}$ e-cm, probably not far from the upper limit. Other supersymmetric approaches predict about 20–1000 times larger than this, so it can distinguish approaches. Confirming this significant suppression is a major test for the compactified M-theory.

The current experimental result is $d_e \leqslant 9 \times 10^{-29}$ e-cm, from the ACME collaboration. These experiments are continuously being improved, so our prediction level could be achieved on a time scale to provide information before colliders.

The compactified M-theory prediction occurs for an important reason. It turns out that the M-theory prediction is that all the Lagrangian terms are real, while an EDM requires a complex theory, so the EDM is zero at the string scale. Running down to the electroweak scale generates a complex result, but a calculable one, since it arises from the CKM phase.

The paper referred to contains predictions for the neutron and Hg EDMs as well.

Reference

[1] Ellis S and Kane G 2016 Theoretical prediction and impact of fundamental electric dipole moments *J. High Energy Phys.* JHEP01(2016)077 (arXiv: 1405.7719)

IOP Publishing

String Theory and the Real World (Second Edition)
The visible sector
Gordon Kane

Chapter 16

Future colliders?

We need colliders, not ideas. Arguably **we have the concepts we need to construct and comprehend the final theory, but we need another, higher energy collider to confirm that we know how the Hierarchy problem is solved. Without another collider we are giving up the age-old search for understanding the world we live in. No amount of astrophysics or cosmology or precision measurements or luminosity frontier can tell us essential data about the Hierarchy problem.**

The Large Hadron Collider (LHC), today's largest collider, at the European CERN Laboratory near Geneva, has a total energy for collisions of about seven times the Fermilab collider. Reasonable improvements in magnet technology would probably allow a similar factor of seven new collider with a ring a few times larger than the LHC one. **The best arguments we can make today suggest less increase in collision energy is needed—perhaps only 2.5 or 3 times increase in collision energy over that of the LHC could produce superpartners.**

As I write, **China is also considering building such a facility. It would be a powerful stimulus to high tech discovery because it will be solving frontier technical problems to get the needed data and analyze it. It would also have a powerful cultural impact, bringing young people to science, and making China the world leader in this fundamental area.**

Europe is also considering building it, connected to CERN. Because of the historical situation after World War II, funding was limited, and the main countries of Europe did not trust each other. They arranged for the contributions from each country to be a treaty obligation, which is much harder not to pay. In addition they indexed the contributions to inflation. Thus **CERN has effectively a guaranteed permanent budget. It has carried out studies of future colliders.** CERN is already committed to upgrading the LHC intensity, which will require a shutdown of perhaps two years plus several years running time. That follows a run of several years starting in mid-2021 or 2022. At that stage, probably a decade, an energy upgrade, or a new tunnel that allows a larger radius could be dug.

We have learned that funding facilities vs strengthening the fabric of society is not a zero-sum activity. That's clear when investments in colliders are compared with other spending categories. We also learned that when a new large U.S. collider (the SuperConducting Supercollider, SSC) was canceled in 1993, and the funds did not enrich other anticipated social or scientific activities. Maybe the strongest reason to build a frontier scientific facility, beyond what exists, is that it attracts young people to science. Then they spread out—about half of CERN PhDs move to fields outside of particle physics, many via startups.

From an economic point of view we have learned to think of colliders as investments. Earlier colliders led to huge technological spinoffs—the World Wide Web; tens of thousands of magnets and accelerators; imaging devices; better chips; and so on. Their economic impact was far greater than the cost of building and operating the collider. That's no accident—it is already happening as people think about how to make the next collider, and deal with the huge amount of data it will produce. **Frontier facilities are a better investment than more of the existing facilities since the most challenging developments are needed to build new facilities.**

The world we see is elegantly described at the level of particle physics and cosmology by the so-called 'Standard Model'. After it was formulated in the 1970s, it dramatically predicted several new particles. A *new collider* was built at CERN, and soon found two of them, the quanta of the electroweak force (W's and Z). The top quark and the Higgs boson were still missing.

A *new collider* with several times larger collision energy was built at Fermi National Accelerator Laboratory near Chicago and consequently the top quark was discovered in the late 1990s. Then another *new collider* was built at CERN about seven times larger than the Fermilab one, the LHC, and the Higgs boson was found there. Without discovering the Higgs boson, most people wouldn't believe such an amazing object existed.

Compactified string theory and M-theory predict several possible discoveries at a *new collider* a few times LHC energies. **The scientific case is strong as summarized here**. That's what the scientists need in order to improve understanding and maintain intellectual growth. Weakly interacting particles with tiny masses *are* something one looks for with large colliders—that's how to produce them.

How massive are superpartners? In recent years people have constructed realistic models based on compactified string/M-theories. If we're looking for predictions for colliders they have to be rather accurate. Models give gluinos (which have a large cross section and good signatures) being heavier than about 1.5 TeV, and lighter than about 5 TeV. We should not have expected gluinos lighter than 1.5 TeV. For realistic gluino signatures the LHC data currently excludes about 1.7 TeV and lighter gluinos. Larger numbers can be quoted, but not for realistic models.

The U.S. funding structure will prevent any future construction of a higher energy facility in the U.S. That's because it would cost some billions of dollars, perhaps needing a billion dollars a year or so. The total cost is not the problem. As we've learned that's an investment. But congress would do the funding a year at a time, so ordering equipment and hiring that require long-term commitments is not possible. The cost should be calculated by setting a price and then paying over several years, but instead it is done with a new appropriation annually. And some years the appropriation might be low, leading to costly delays. **The physics argument for needing the higher energy collider can be questioned by people who don't understand the need to solve the Hierarchy problem, allowing lower annual budgets. Only Europe (CERN) and China can evade these difficulties.**

The needed collider may be designed as a staged facility. A tunnel larger than that of the LHC, with an initial electron–positron collider designed to study Higgs physics, followed by a proton–proton collider with stronger magnets would be fine (except for those of us eager to have answers, since it would not take data for many years).

IOP Publishing

String Theory and the Real World (Second Edition)
The visible sector
Gordon Kane

Chapter 17

Three families, quark mass hierarchies and splittings

A number of people have **used the adjoint representation of a single E₈ singularity in a visible sector to work toward three families of quarks and leptons. We do this in a new way, breaking E₈ to the Standard Model via deformations and geometric engineering following the technique of Katz and Morrison [1].** We work in the compactified M-theory framework. Past work on E₈ via other ways of symmetry breaking has run into problems. **We hope to understand the origin of flavors and three families, and the values of quark and lepton masses.** These are of course all internal symmetries, not including Lorentz transformations, so that one cannot change flavor by changing position or motion.

The initial effort is partially successful. **We can see three families and the Hierarchy of quark and charged lepton masses emerge. We can see the isospin breaking that makes the SU(2) doublets such as top and bottom, up and down, electron and electron neutrino have their mass splittings, and the Hierarchy of family masses. The amounts are controlled by deformation parameters that are effectively moduli. We can calculate the values of the deformation moduli that lead to the hierarchies, and realistic values for the masses.** In earlier work we derived the result that $\tan \beta \sim 7$, which is used in relating the masses.

Ideally we would be able to derive the values at which the deformation moduli are stabilized, and predict the masses, but we cannot do those things yet. We are in the process of adding neutrino masses. It would be good to add global G_2 constraints.

We are able to get some important mass values. We work with high scale Yukawa couplings. **The top quark has a Yukawa coupling of order one. The up quark satisfies $m_{up} + m_e < m_{down}$ (ignoring an electromagnetic contribution) so that protons will be stable rather than neutrons, allowing hydrogen atoms.** We can derive the conditions in the underlying theory for these results but not yet show they can be uniquely derived. **That the theory allows them in a UV complete theory is encouraging.**

doi:10.1088/978-0-7503-3583-6ch17

Reference

[1] Katz S and Morrison D 1992 Gorenstein threefold singularities with small resolutions via invariant theory for Weyl groups *J. Alg. Geom.* **1** 449–530 (arXiv: alg-geom/9202002)

IOP Publishing

String Theory and the Real World (Second Edition)
The visible sector
Gordon Kane

Chapter 18

How much can we understand?

Includes extracts from *Supersymmetry: Unveiling the Ultimate Laws of Nature* by Gordon Kane, copyright 2000. Reprinted by permission of Basic Books, an imprint of Hachette Book Group, Inc.

Suppose that superpartners are indeed detected and studied experimentally soon. Suppose that from their properties we are led to a string/M-theory whose four-dimensional form explains the features we think are essential for understanding the world of particles and cosmology. How much further can we hope to go in understanding where the laws of nature and the Universe came from—are there limits? Even if we can eventually understand the Universe, is it premature to hope to do so soon? Now I'm talking about the hard questions. Could the laws of nature be different? Once we say the rules are quantum theory and relativity, and string/M-theory, we've seen we can make progress, but why those rules?

As always with science, we can't know what we can understand until we get there, so we will only be certain of the answers to these questions if we succeed. But we can look at the arguments against the possibility of learning and testing the theory more deeply and see how good they seem to be. The outcome is that skepticism is not backed up by sound arguments.

First consider whether we can hope to discover the final theory. We think we have done that by solving the problems in the list, but maybe we're fooling ourselves. Remember, our current goal is to *understand our* world. Both italicized words are crucial. People have raised a number of issues meant to suggest that we will not achieve that understanding. One argument is that there is much that we cannot know about the world, and science has added to that list. For example, because of the finite speed of light and information, we cannot know what is happening right now on Mars, or halfway across the Universe. We cannot know the position and momentum of a particle simultaneously to better than some accuracy set by the uncertainty principle. But learning that there are things we cannot know (as in the above examples) is actually part of that understanding, and in no way implies that we cannot understand how the Universe works and why it is the way it is. We cannot know all possible chemical molecules, but we can fully understand the principles that govern the formation and behavior of all possible molecules. Understanding whether there are many possible universes can be studied in parallel with understanding our world—both are important.

Another argument is based on Gödel's incompleteness theorem. That is an astonishing and elegant mathematical result proved by Kurt Gödel in the 1930s. For our purposes it basically states that in any mathematical theory that is interesting for our purposes, there are true results that cannot be mathematically proven to be true, and also that you can't prove the consistency of a mathematical system from within that system. For mathematicians that has profound implications. A number of people have worried that it also meant that physicists would be unable to show that the final theory applied to the world even if it did.

But that is not how science works. There are two important differences. First, we do not need to prove all possible theorems, nor do we have to prove the consistency of the whole system. We already have our world, and we know it is described by consistent laws—otherwise it would fall apart. If the equations for the stability of atoms changed with time, or were inconsistent with those for the forces, atoms would not keep existing and forming the world. Indeed, it is often remarked that it is 'amazing' that our world is comprehensible scientifically. But is that really surprising? Our world must behave according to mathematical regularities if it is to exist for some time. Once that is the case, we can learn what the regularities are.

Second, scientific results are never proven to be true. 'Proof' is for mathematical theorems. At a certain stage of research, the evidence for a given physics result becomes so strong it is accepted by those who understand it, and by others who trust them. Every result depends on certain parameters (such as distance or speed or others). If the evidence comes only from a limited range of those parameters, the result may or may not change when it is extrapolated outside that range; recall the earlier discussions. For example, the laws of gravity have not been tested for some distances smaller than a fraction of a millimeter. Experiments, motivated by some ideas from string/M-theory and extra dimensions, are underway to find out if the form of the gravitational force changes at smaller distances. On the other hand, our descriptions of all forces have now been tested for all speeds, from rest to the maximum possible speed (of light), and so there will not be further modification there.

In addition, as we have seen, accepted scientific results form a coherent structure with many implications that strengthen our confidence in them. If one part is modified the whole structure might change and contradict results that are only related when there is a theory. This is a very important guide.

The results of science lead to our surest knowledge about our world because of the process used to obtain them, including improved experimentation and consensus of informed workers. The bottom line is that Gödel's theorem is simply not relevant to whether we can understand how our particular world works and why it is the way it is.

Yet another concern arises from a feeling of humility in the face of the awesome size and complexity and beauty of the Universe, from the particles to the cosmos. How can mere humans expect or even hope to understand all of that and how it originated and why it is the way it is? Sometimes this is stated with the comment that it is like expecting a dog to understand quantum theory. Charles Darwin first used that analogy—he wrote 'we feel most deeply that the whole subject is too profound for the human intellect. A dog might as well speculate on the mind of Newton.' It is easy to understand why Darwin would feel that way living as he did before all that has been learned in the past century and a half. We won't know if humans can figure out the final theory unless our efforts to understand either succeed or hit a dead end. The approach of science since it began is simply to try to understand natural phenomena as well as possible, and to see how far we can go. Personally, I think that if a hypothetical dog were able to get data about atoms and ask about how they work, then it could discover quantum theory, so I remain hopeful that we will indeed understand the Universe.

Testing string theory and the final theory

Could we fail to understand the world because we cannot test ideas? This is a more subtle and interesting point. Some people have suggested that because we can never build a collider that can directly probe the Planck scale, we can never test ideas about physics there, or test the primary theory. That is simply wrong, as we showed earlier. We did not have to be there when the dinosaurs become extinct to figure out how that happened—again there are clues that let us unravel the mystery. We do not have to travel near the speed of light to figure out and confirm that it is the maximum speed at which we can travel. There are always relics that will help us test ideas. Finding those clues and relics requires dedicated effort by talented scientists, and financial long-term commitments from governments, or billionaires. It doesn't happen by accident; first ideas and hypotheses have to be formulated, and then tests emerge. People argued in the past that it was not possible to demonstrate that atoms existed, or that neutrinos existed. As the saying goes, 'impossible just means it takes a little longer.'

Similarly, string/M-theory can address a number of questions that the Standard Model cannot address, such as the number of families of quarks and leptons, the values of the masses of the quarks and leptons, which phenomena should show CP violation, and whether protons decay. If string/M-theory incorporates quantum theory and gravity and the Standard Model forces into one description in a consistent way, and also explains why there are three families of quarks and leptons and calculates some of their masses from formulas that have no adjustable parameters, we will surely have the understanding of our world that we hope for. The problem is that the calculations needed to be sure string/M-theory does all of that are very difficult, and they depend on how the small dimensions curl up and on which of many solutions nature actually selected. So even if string/M-theory were indeed right we would not know it until people are able to do the calculations, and confirm that string/M-theory does indeed explain the unsolved problems. The answers proposed in this book for the final theory list have to be checked. The problem is partly psychological. For example, for some time the development of a quantum theory of electromagnetism was hindered because a number of calculations of observables gave infinite results when they shouldn't. Then in 1947 a measurement was reported for one of those observables, and soon after theorists figured out how to make the calculations finite and meaningful and got the right answer. What changed was they took it more seriously when there was data, and they had a definite experimental answer to let them know if the result of their calculation was right.

With string/M-theory the question is whether it will be taken seriously enough so talented people will focus on the calculations that test the ideas even when that means investing months or years in work that might not pan out. I think that is probably the situation with the cosmological constant problem. Hopefully the situation will be similar with the Higgs boson mass now measured and calculable in compactified theories, so it can push theorists to understand better how to derive more predictions.

The situation will be similar for the final theory. If a candidate theory allows us to understand Higgs physics, parity violation, inflation, the Hierarchy problem, three families and more we will not have much doubt it is a strong candidate for the final theory. There may be additional tests too, though until we get there we won't know what the tests are. Many of the tests of any idea only emerge after the idea is formulated and its implications studied (as happened with the tests of the Big Bang, or the prediction of radio waves from Maxwell's equations, for example). The key point is that since we can describe some possible tests of the final theory, it is clear that any arguments that the final theory is in principle untestable are not valid. In practice it may be hard to do the same tests, and we may have to rely on the tests with classes of compactified theories as the best we can do for some.

Practical limits?

Another possible limit to our ability to understand nature may arise because society is unwilling to provide the funding and the commitment to do the basic research needed to test the ideas. As we have argued in this book, if there is supersymmetry at the collider scale then there is a good chance that the facilities whose upgrade or construction is underway or planned, combined with experiments to study dark matter and proton decay and CP violation and neutrino masses, and with cosmological data, will provide us with the information needed to formulate and test the final theory. Even if these facilities are available there are economic risks, because to build and operate these frontier facilities a number of talented and committed people are needed. If the funding comes too slowly these people will be forced to leave the field, as many had to do when Congress terminated the supercollider project. The people who do this frontier research can only be based at top research universities and a few national laboratories. If those institutions reduce their commitment there will not be positions for the theorists and experimenters, or places to train the bright young people who want to learn how and why the Universe works, or teachers to train and inspire them.

It is important and interesting to examine the justifications for the funding. In recent years it has been fully understood and documented that much of our economy is based on earlier funding of scientific research. One might think it is the results of research that drive the economy, and they do, but it is not only the results. Fascination with what has been learned from science, and the way science can lead to understanding our world, initially attracts young people to a career in scientific or engineering research. What they end up working on after their education can be very different from what brought them in initially. Many of them see an opportunity to develop products or information technology. The ways that understanding gained from science enriches our culture and our view of our relation to the Universe, and the impact of basic research on young people, are probably the two strongest justifications for any society to support basic research well.

But even apart from these benefits, basic research more than pays society back for its cost through the mechanism known as 'spinoffs'. The results of research about Higgs bosons or superpartners or dark matter are not likely to lead to products that directly affect the economy. But because these research areas are probing new frontiers, scientists *must* develop new techniques, and these new techniques invariably lead to new industries. The most spectacular example is the World Wide Web, developed at CERN in order to find new ways for international collaborators to handle data from the LEP collider. The same is not true for constructing the second or third facility of a given type, since then the new developments have often been made.

Another example is accelerators, which were invented to probe more deeply into nuclei and protons, and continuously developed to do more particle physics. They now are used to study materials and matter in many ways, providing knowledge about how to make stronger and safer materials, to find the structure of viruses, and much, much more. There are over 30 000 thousand accelerators in use in the world today, and less than a few per cent are used for particle physics research. Accelerators have so many medical-related uses that the National Institutes of Health have increased funding to support not only people who use the accelerators but even the accelerators themselves. The list of spinoffs, and of associated startup companies that can initially survive because of the guaranteed market provided by particle physics labs and experiments, is very long, and shows convincingly that even if the intellectual interest of the results of the research is not included, funding particle physics and cosmology is an investment that brings large economic returns to society. It is often stated that most new U.S. jobs come from Research and Development, and the majority of startups emerge from frontier research. A useful web site for more information on these topics is http://www.interactions.org/cms/? pid=1003378. It would be good if more of the people who control the funding in Washington and at our research universities appreciated this.

The role of extra dimensions

We saw that the Standard Model, formulated in four dimensions, gave us a descriptive understanding of the particles and forces, but not a 'why' understanding. That remained so for the supersymmetric extension of the Standard Model. But compactified string/M-theories could address 'why' the particles and forces and even the rules were the way they were. Having the extra dimensions, and formulating the underlying theory including their role, may allow us to go beyond a descriptive understanding to a deeper one. The extra dimensions are not a complication but an essential aspect of gaining understanding. The value of the Higgs boson mass provides an example. In the Standard Model the value of the Higgs boson mass cannot be calculated at all. In the supersymmetric extension of the Standard Model the mass can be estimated to be larger than about tens of proton masses and smaller than about two hundred proton masses, better but not what we would hope for. In compactified string/M-theories the Higgs boson mass can be calculated with an uncertainty of less than a few per cent, just as we would hope would be the case if we had a deep understanding.

Sometimes people comment or argue that the theory has gotten very complicated, with a doubling of the number of particles and with extra dimensions. There is a vague hope that correct theories should be 'simple', without fixes and add-ons if problems occur, without epicycles. An 'Occam's razor' argument that simpler theories are more likely to be true is sometimes used. In fact we would argue that the compactified string/M-theories are the simplest theories that could possibly encompass and integrate all the phenomena of the natural world into one coherent mathematical theory. The compactified string/M-theory that is emerging is as simple as it could be.

Implications?

Suppose that eventually we do indeed figure out what the final theory is, and are able to test it so well that most of us are convinced that we indeed understand how our Universe works and why it is that way, and the underlying theory framework requires (or does not) a multiverse. What might that imply? Although humans began to wonder about the Universe perhaps 40 000 years ago, and started a more systematic quest to understand it about 2600 years ago in Greece, until about 1600— a mere 400 years ago—there was only a little descriptive progress. Then modern science began, and today we have started (and perhaps nearly finished) research on the final theory itself. Before the 1970s we did not know how the world worked, and now, to a large extent, we do. Perhaps in a few decades or less we will reach the end of this quest. Our Universe has properties consistent with having humans in it, though it is indifferent to whether they are there, or whether they understand it. For us, and we hope for many people, achieving an understanding of why there is a universe and why it is the way it is will be a source of immense pride and dignity and meaning in the face of that indifference.

Many people, and many scientists, have said that we will not reach that end, that there will always be new questions, that each discovery will lead to additional questions. Why should that be so? There is no known reason why the quest should not end. Sometimes the analogy with exploration of the surface of the Earth is helpful—for many centuries there was more to explore, and then one day we were essentially done. There have been a few occasions when someone said that there was nothing new to learn, but if you actually examine the concerns of leading scientists since the 1860s you find the active scientists always knew that was wrong, and addressed many puzzles. The situation truly is different today. Now we have tested our description back to the beginning of the Universe, out to the edge of the Universe, and down to the fundamental constituents of matter. This doesn't guarantee we will achieve total understanding, but it does demonstrate that the historical analogies need not be relevant to the argument.

Yes, it is possible that science in the direction of effective theories with fewer inputs could end, not because we couldn't get all the way to the final theory and no inputs but the Planck scales, but because we did. It would mean we knew the equations and the laws, but we would still be far from knowing all the solutions, and what might emerge from them.

IOP Publishing

String Theory and the Real World (Second Edition)
The visible sector
Gordon Kane

Appendix A

M-theory calculations

Hopefully the main body of text has been fairly comprehensible to most readers. There's a price to pay for that, making a lot of claims instead of derivations. Compactified M-theory results are not widely known, so it seems appropriate to sketch out the derivations in such a way that readers can find them if they want. The results are published, but for convenience I'll just give the arXiv number since that's the easiest way to find anything. I'll only mention a few results needed for us. With apologies, I'll also not do a complete referencing job, but only list what is helpful for understanding the history.

- Edward Witten presented his discovery of 11-dimensional M-theory in 1995.
- G Papadopoulos and P Townsend quickly found the result that compactifying M-theory on a 7-dimensional manifold of G_2 holonomy left a resulting 4-dimensional high scale relativistic quantum field theory that was automatically supersymmetric, so the supersymmetry did not have to be assumed or put in, and it had four large dimensions.
- Then a postdoc, Bobby Acharya (now Professor at King's College, London), showed that non-Abelian Yang–Mills gauge fields were localized on singular 3-cycles in the 7D manifold (arXiv: hep-th/9812205). These could be interpreted as the Standard Model gauge fields (and their superpartners).
- Next, Acharya and Witten, arXiv: hep-th/0109152, showed that chiral fermions were supported at points with conical singularities on the manifold. These were the quarks and leptons. Thus all the particles of the Standard Model could be identified.
- In 2004 Acharya and Sergei Gukov summarized and reviewed the field in Physics Reports.

At that stage Acharya and I (and PhD students and postdocs) began to work on applying the theory to the real world. We made a few discrete assumptions:
- We would work, for well-motivated reasons, in the fluxless sector, which is technically robust. Basically the idea was that fluxes, which were essentially generalizations of electric and magnetic fields, had dimensions and were

naturally at the Planck scale. Our goal was to get TeV physics, which should be easier if we did not have to overcome Planck-scale physics. Technically this was possible.

- We assumed we could use generic Kähler potentials and generic gauge kinetic functions, which had already been worked out by others. The Kähler potential was $K = -3\ln((4\sqrt[3]{\pi})\,V)$, where $V = \prod_{i=1}^{N} s_i^{a_i}$, with $\sum_{i=1}^{N} a_i = 7/3$. V is the 7D volume and the s_i are moduli [See Beasley and Witten arXiv: hep-th/0203061; Acharya, Denef, Valandro arXiv: hep-th/0502060]. The dependence of the Kähler potential on the moduli is only through the volume.

- The gauge kinetic functions had also been calculated by Lucas and Morris, arXiv: hep-th/0305078. They are $f_k = \sum_i^N N_i^k z_i$, integer linear combinations of all the moduli.

- Include massless visible and hidden sector chiral fermion states Q (with various flavors and colors), which give a superpotential $W = A\phi e^{ib_1 f_1}$, with $b_1 = 2\pi/c_1$ and c_1 the dual coxeter number of the relevant hidden sector gauge group. The A's are constants of order unity. [See Seiberg arXiv: hep-th/9402044, arXiv: hep-ph/9309335; Lebedev, Nilles, Ratz arXiv: hep-th/0603047.] ϕ is a complex effective meson field, $\phi = \det(Q\tilde{Q})^{1/2} = \phi_0\, e^{i\theta}$. If we approximate the superpotential by two terms we can get useful solutions with good properties such as being in the supergravity regime, simple enough to do most calculations semi-analytically and also numerically, etc.

- Imagine expanding the exponentials. Then all terms get interactions, and every modulus has a potential to stabilize it. The A_k are constants of order unity. The microscopic constants A_k, b_k, a_i, N^k_i are determined for a given G_2 manifold.

- There are hidden sectors that lead to breaking supersymmetry, de Sitter vacuum, dark matter, baryogenesis, LSP decay and other phenomena.

- Most of the derivations are in Acharya, Kane et al, arXiv: hep-th/0701034.

- We assume that the needed singular G_2 manifolds exist. There is considerable progress about this because of the Simons Foundation funding a study of manifolds of special holonomy initially for four years and nearly $10 million, and now extended for three years. G_2 is the main manifold of special holonomy studied.

- The little Hierarchy problem may be solved. The electroweak symmetry breaking condition requires the solution to give M_Z, whose value is about 91 GeV, from an equation whose individual terms are all of order one to two TeV. Each term is subject to radiative corrections, so the solution is very sensitive to details. There is a solution in the right range, but there is extreme sensitivity to whether it is indeed a true solution.

We see here that the definition of a particle is a massless mode of an excitation. It will be charged under a number of symmetry groups, i.e. carry electroweak charge, color charge, various U(1) charges, and less familiar ones such as E_6 charges.

Acharya and I (with students and postdocs) started then. We focused on Higgs and LHC physics. Indeed, we showed that in M-theory supersymmetry automatically was spontaneously broken via gaugino condensation, was gravity mediated, and simultaneously all moduli stabilized in a de Sitter vacuum. The scalar potential is then calculated and minimized. Note that this is not an effective theory, since all corrections are volume suppressed—corrections to Kähler potential, string loop, α' etc. We calculated the supersymmetry soft breaking Lagrangian, which led to radiative electroweak symmetry breaking, Higgs physics, precise ratio of M_h/M_Z and Higgs decays, in the decoupling sector, approximate gaugino masses, heavy squark and heavy M_{hu} and M_{hd}, and surprisingly very suppressed electric dipole moments. The only dimensionful input is the Planck scale. The gravitino mass is calculated to be of order 35 TeV. Gaugino masses are suppressed and are of average order 1 TeV. The ratio of M_h to M_Z is correctly predicted. The only dimensionful input is the Planck scale. We will see that the other scales of the microscopic particle world are calculable. It should be emphasized that this is a theory that is by construction UV complete. It solves the Hierarchy problem. It cannot be in the swampland. It does not have free parameters. It may seem complicated, but it is arguably the simplest theory that could describe the phenomena of our world.

IOP Publishing

String Theory and the Real World (Second Edition)
The visible sector
Gordon Kane

Appendix B

Kaluza–Klein extra dimensions

It turns out that the extra dimensions can play an exciting role. The argument is a little technical, but one can see the general idea without understanding the details. We of course have three large space dimensions and one time dimension, so six (or seven) curled up ones. A century ago theorists started thinking about extra dimensions. Theodor Kaluza wondered what would happen if he imagined there was just one extra dimension. He was working shortly after Einstein introduced general relativity. In general relativity one first writes the 'metric', which describes the geometry of space–time. Let's call the metric of the five-dimensional world g_{ab}, a, $b = 0,1,2,3,4$, and the metric of our four-dimensional world $g_{\mu\rho}$, $\mu,\rho = 0,1,2,3$.

Then Kaluza showed one could write g_{ab} in a matrix form,

$$g_{a,\,b} = \begin{pmatrix} g_{\mu\nu} & A_\mu \\ A_\mu & \varphi \end{pmatrix}$$

where A_μ can be interpreted as the vector potential of electromagnetism, $g_{\mu\rho}$ the four-dimensional metric, and φ is a scalar we can ignore. The exciting thing is that the extra dimension can be interpreted as a force, analogous to the electromagnetic force. The details are not correct, but the concept was exciting. A few years later, Oskar Klein, then an assistant professor in the University of Michigan physics department, suggested that the extra space dimension could not be seen because it was curled up with too small a size for us to see it. The Kaluza–Klein theory didn't work in detail for describing our world, but it pointed in the right direction.

To describe our world we have to take account of the need for a theory with ten dimensions while we only experience four dimensions. Projecting the full theory onto ours is called 'compactifying'. Just as an atom falls into its ground state, our world will fall into its ground state, which is usually called the vacuum, and is usually compactified. The remarkable thing is that a compactified ten-dimensional world can contain our world including the forces of the Standard Model. We live in the ground state of a compactified string/M-theory. Because we have included gravity in the theory it also has an ultraviolet completion, it is 'UV complete'.

doi:10.1088/978-0-7503-3583-6ch20

The Kaluza–Klein idea, that the extra dimensions turn effectively into forces in the compactified theory, encourages us to think that compactified string/M-theories are the way to make a UV complete theory of our world, and hopefully solve the Hierarchy problem.

IOP Publishing

String Theory and the Real World (Second Edition)
The visible sector
Gordon Kane

Appendix C

Significant results for which we don't need to know the mathematical details of manifolds

There are a surprising number of important results that hold regardless of the details of manifolds, once we work them out. No systematic effort was made to find additional such results.

- The stabilization of all moduli;
- The gravity mediation of supersymmetry breaking;
- Scalars will be heavy, of order the gravitino mass (35 TeV). This includes squarks, sleptons, Higgs soft-breaking masses;
- Radiative electroweak symmetry breaking is generic, leading to the ratio of the Higgs mass to the Z mass, and the Higgs decay branching ratios;
- The cosmological history is matter dominated, from shortly after inflation ends to shortly before nucleosynthesis;
- There is a de Sitter vacuum;
- The inflaton is to a good approximation the overall volume modulus;
- The lightest superpartner will decay to hidden sector matter;
- Gaugino masses are light, about a TeV on average, with charginos and neutralinos about half a TeV and gluinos about 1.5 TeV or heavier;
- Electric dipole moments are surprisingly small.

IOP Publishing

String Theory and the Real World (Second Edition)
The visible sector
Gordon Kane

Appendix D

De Sittter vacuum

Different vacua often carry the name of the first person to study a universe with only the energy density of such a vacuum. We live in a de Sitter vacuum, with an accelerating expansion rate.

Unfortunately, it has been 'known' for a long time that universes filled with moduli lead to an anti-de Sitter universe, and compactified string theory ones were included in that category, apparently ruling out all compactified ones.

Under the assumptions usually made for that result it is indeed correct. But the assumptions include neglecting hidden sector charged matter, and once the hidden sectors are included the vacuum indeed becomes a de Sitter one.

This emerged in the calculations carried out in Acharya *et al*, arXiv: hep-th/0701034v3.

We can sketch the result qualitatively. The potential can be written

$$V = e^{K/2}(F^*{}_{\text{Hidden}}F_{\text{Hidden}} + F^*{}_{\text{Moduli}}F_{\text{Moduli}} - 3\,|W|^2)$$

where the last term is small, and the moduli F-terms are not large enough to overcome the last term, but the hidden sector moduli F-terms generically lead to a large positive contribution that gives a (positive) de Sitter potential for typical compactifications.

There is a large literature on this topic, initiated by Cumrun Vafa in modern times, trying to get a de Sitter vacuum with quantum corrections, but ignoring hidden sector physics (which in general also are quantum corrections). Since the hidden sector effects suffice, I will not review other attempts. Many of them do not include the hidden sector matter. Generically compactified theories have hidden sectors and de Sitter vacua.

IOP Publishing

String Theory and the Real World (Second Edition)
The visible sector
Gordon Kane

Glossary

Symbols

e; $\tilde{e}$	electron; selectron.
μ; $\tilde{\mu}$	muon; smuon.
τ; $\tilde{\tau}$	tau; stau.
ν; $\tilde{\nu}$	neutrino; sneutrino (there are separate sneutrinos for electron, muon, and tau, sometimes written with subscripts).
u, d, c, s, t, b	quarks (up, down, charm, strange, top, bottom).
$\tilde{u}, \tilde{d}, \tilde{c}, \tilde{s}, \tilde{t}, \tilde{b}$	squarks (up squark, down squark, charm squark, strange squark, top squark (sometimes 'stop'), bottom squark ('sbotto')).
W; $\tilde{W}$	W boson; Wino.
Z; $\tilde{Z}$	Z boson; Zino.
G	Newton's constant, that determines the strength of the gravitational force.
h	Planck's constant, that determines the size of quanta of energy and of quantum.
c	Einstein's constant the speed of light in vacuum (or charm quark, depending on context).
Powers of ten	10^{-6} = one millionth, 10^{-2} = 1/100 (one/hundredth), 10^0 = 1, 10^1 = 10, 10^2 = 100, 10^3 = 1000, 10^6 = one million, 10^9 = one billion.

Acronyms

CDF	Detector at Fermilab.
CERN	Centre Européan Recherche Nucléaire, Geneva, Switzerland.
D0	Detector at Fermilab.
Fermilab	Fermi National Accelerator Laboratory, Chicago, IL.
LEP	Large Electron–Positron Collider, at CERN.
LHC	Large proton–proton collider at CERN.

NLC The 'next' linear collider—the possibility of proposing such a collider is under study by several countries and laboratories. There are discussions about making it an international facility; it would be an electron–positron collider with energy a few times that of the SLAC SLC.

SLAC Stanford Linear Accelerator Center, Palo Alto, CA.

SLC Linear electron–positron collider at SLAC.

ACCELERATOR Accelerators are machines that use electric fields to accelerate electrically charged particles (electrons, protons, and their antiparticles) to higher energies. If accelerators are linear they need to be very long to achieve the desired energies, so some use magnets to bend the particles around and back to the starting point, giving them a little extra energy each time around.

ANTIMATTER Every particle has an associated antiparticle, another particle with the same mass but with all charges opposite. If a particle has no charges, e.g. the photon, it is its own antiparticle. Often the antiparticle is denoted by writing a bar over the particle name, e.g. $\bar{e}$ for the electron antiparticle (also called the positron).

ATOM An atom has a nucleus surrounded by electrons, bound together by the electromagnetic force. Ninety-two different atoms occur naturally, making ninety-two different chemical elements, with nuclei having 1–92 protons. The atoms are electrically neutral. The diameter of an atom is about 10 000 times larger than the diameter of its nucleus.

AXION Axions are a hypothetical particle. They were invented to solve a problem for QCD. The QCD Lagrangian is allowed to have a term that violated the CP symmetry. Then the strong interactions should show CP violation, but they do not, to an accuracy of about a part in 10 billion. Assume the vacuum contains a (pseudoscalar) field like the electromagnetic field, that interacts with the gluon field in a way that just cancels the CP violating gluon interactions. Axions are the quanta of that field. Such a particle would have a definite mass and interaction strength. It turns out such a particle could be the dark matter. The meaning has been generalized to apply to any pseudoscalar field. String/M-theories have a large number of axions, probably one associated with each modulus.

b-FACTORY

A b-factory is a facility designed to produce and detect large numbers of b quarks, at least 100 million a year. Planned b-factories are electron–positron colliders, but a proton collider could also be used if an appropriate detector could be made. The main goal of b-factories is to study CP violation.

BARYON

A baryon is a composite particle made of three quarks, any three of the six. Protons and neutrons are baryons.

BARYON ASYMMETRY

See MATTER ASYMMETRY.

BEAMS

One way to learn more about particles is to collide them with one another and see what happens. Beams of electrons and protons can be made by knocking apart hydrogen atoms and applying electric fields. Positrons and antiprotons don't exist naturally, because they annihilate as soon as they encounter an electron or proton; they can be made by hitting a target with energetic protons or electrons, and then collected by placing magnets after the target, arranged so as to bend each kind of particle in a different path. Then bunches of them are accelerated to higher energies. When a particle hits a target, every kind of particle is made with a certain probability, so other beams of particles can also be made (neutrons, muons, kaons, neutrinos, etc) by judicious arrangements of magnets and material.

BIG BANG

Several strong kinds of evidence imply that our Universe began as a tiny dense gas of particles which has been expanding since, i.e. our Universe began in a 'hot Big Bang'.

BLACK HOLE

Black holes form whenever sufficient matter is packed into a small enough space so that the resulting gravitational force at the surface is strong enough to prevent anything, including light, from escaping. They can be microscopic in size, or formed from billions of stars. They can be detected from their indirect effects on nearby matter. The theoretical study of their properties can greatly clarify our understanding of basic questions.

BOSON

Bosons are any particles that carry an integer unit of spin (0, 1, ...). They have different properties from particles with half a unit of spin (fermions). In particle physics 'boson' has a more specific usage—bosons (photons, gluons, W's, and Z) are particles that are the quanta of the electromagnetic, strong, and weak fields. They transmit the effects of the forces between quarks, leptons, and themselves. Higgs bosons are quanta of a Higgs field.

CHARGE—ELECTRIC, COLOR, WEAK

Each particle can carry several kinds of charge that determine how it interacts with others. Electric charge is familiar to us in everyday life. Particles can have positive or negative electric charge, or none. Color charge and weak charge are not familiar because their effects can only be felt at distances smaller than the size of a nucleus. Color charge and strong charge are the same thing. A particle cannot have random amounts of charge; only certain discrete amounts are allowed. The extent to which a particle feels each force is proportional to its associated charge. Quarks and gluons carry color charge; quarks and leptons, and W and Z bosons, carry weak charge.

CHEMICAL ELEMENTS

Ninety-two different stable nuclei can be formed from neutrons and protons bound together. Each forms atoms by binding as many electrons to the nucleus as it has protons (so the nucleus is electrically neutral), giving 92 different atoms. These atoms are the smallest recognizable units of the 92 chemical elements.

COLD DARK MATTER

Particle physics theories that extend the Standard Model often predict the existence of new, stable particles that were present in the early Universe and survive to the present, making up a large fraction of the matter of the Universe. These particles interact weakly and they are usually massive so they move slowly—they are cold. An example of such a particle is the lightest supersymmetric partner if it is stable. Astronomers have evidence from the motion of galaxies, and from the large scale structure of the Universe, that cold dark matter exists.

COLLIDER

A collider is made by accelerating beams of particles and causing them to collide. The energy of the colliding beams can provide much more energy that can be used to make new particles than if the beams hit stationary targets. Colliders have two challenges, getting to larger energies, and getting to higher intensities.

COLOR CHARGE

See CHARGE. Color charge means the same thing as strong charge.

COLOR FIELD — Any particle carrying color or strong charge has an associated color or strong field around it. Any other particle carrying color charge feels that field and interacts with the first particle.

COLOR FORCE — The force between two particles carrying color charge. The color force (or strong force) binds quarks into protons and neutrons. The residual color force outside protons and neutrons is the nuclear force that binds protons and neutrons into nuclei. The color force is mediated by the exchange of gluons.

COMPACTIFICATION — Compactification for us is projecting a theory in some number of dimensions to a theory with some large dimensions and a Planck-scale size manifold of the remaining small dimensions, for example M-theory in 11 dimensions to a 4D Minkowski space and a 7D Planck-scale size manifold, or SU(3) to U(1) × SU(2). In order to have a UV complete theory we need to start in 10D or 11D, but we live and do experiments in 4D so it is necessary to compactify to 4D.

COMPACTIFIED STRING THEORY — Compactification starting with one of the 10D string theories.

COMPACTIFIED M-THEORY — Compactification starting with 11D M-theory.

COMPOSITE — Any object made of other objects is composite, as are atoms, nuclei, and protons. If quarks and leptons had followed the historical trend that each level of matter turned out to be composites made of smaller constituents, experiments should already have shown evidence of their compositeness. That, combined with theoretical arguments, strongly suggests quarks and leptons may be the ultimate constituents of matter, the indivisible 'atoms' of the Greeks.

CONSTITUENTS — Any objects that are bound together to make larger objects. See COMPOSITE. For example, atoms are constituents of molecules, nuclei are constituents of atoms, etc.

COSMIC RAYS — Protons and some nuclei that are ejected from stars, especially supernova explosions, move throughout all space. They impinge on the earth from all directions, and are called 'cosmic rays'. They normally collide with nuclei of atoms in the atmosphere, producing more 'secondary' particles, mainly electrons, muons, pions, etc. A number of cosmic ray particles go through each of us every second, and they can interact in detectors and mimic signals of previously undetected particles, so experimental equipment must shield against them or be able to recognize them so they can be discounted as signals of new physics.

COSMOLOGY

Cosmology is the study of the Universe as a whole, its properties, and its origin.

COSMOLOGICAL CONSTANT

The cosmological constant is the name of a possible term in the equations that describe the Universe. If it is not zero, then it implies there is a force that is slowly increasing the expansion rate of the Universe. There are two puzzles about the cosmological constant. First, the observed value seems to be far smaller than estimates would imply, and second, recent data suggests it is not exactly zero, so an explanation is needed for why it has a particular non-zero value.

CP VIOLATION

Interactions of quarks, leptons, and bosons are normally invariant under a symmetry operation called 'CP', the combined operations of 'Parity' and 'Charge Conjugation'. A small violation of this invariance is observed for some interactions, which may have important implications and be an important clue to deeper understanding of how nature works.

DARK MATTER

Particle physics theories that extend the Standard Model predict several forms of matter that may exist in large quantities throughout the Universe, and make up most of the matter of the Universe. Some move slowly and are called 'cold dark matter', others move rapidly and are 'hot dark matter'. Study of the motions of galaxies and of the formation of clusters of galaxies suggest that such dark matter exists, as do theoretical cosmological arguments based on other data. See COLD DARK MATTER, HOT DARK MATTER.

DECAY

The quarks and leptons and bosons that are the particles of the Standard theory have interactions that allow them to make transitions into one another. Whenever one of them can make transitions into lighter ones, that transition will occur with a certain probability, and we say the heavier one is unstable and has decayed into the lighter ones. Decays are really transitions—the final particles were not contained in the initial one. The initial particle disappears, and the final ones are created. In the Standard Model the up quark, the electron, and the neutrinos do not decay; the other fermions, and the W and Z do decay. All of the superpartners except the lightest one are expected to decay.

DETECTOR

The properties of particles and their interactions are studied by observing their interactions and decays. These observations are done with detectors, which can be thought of as cameras that record information in several ways, not only on film. In order to obtain new results at the forefront of today's questions, detectors have to be very large and usually require the development of new and better technologies. Every particle physics experiment has one or more detectors.

DEUTERON

A deuteron is the second heaviest nucleus (after the lightest, hydrogen, which has a single proton). It is composed of a neutron and a proton bound together by the nuclear force. The deuterium atom has a single electron bound to a deuteron (one electron because there is one proton).

DIRAC EQUATION

The Dirac Equation incorporates the requirements of both quantum theory and special relativity in the description of the behavior of fermions. It requires that fermions have the property called spin, and it predicts the existence of antiparticles. It was written by Paul Dirac in 1928.

DUALITIES—DO NOT COMMUTE WITH COMPACTIFICATION EFFECTIVE THEORY

Each part of the physical world can be described by a sub-theory that applies over a certain distance or energy scale. Such sub-theories are called effective theories. Explanations in a given effective theory can ignore much of the rest of the world. The rest has effects on the part of interest through a few inputs or parameters. Every part of our description of the physical world is an effective theory except possibly the final theory.

ELECTRIC CHARGE

See CHARGE.

ELECTRON

A fundamental particle. The electron has one unit of negative electric charge, and one-half unit of spin. It is a fermion.

ELECTRON COLLIDER (short for electron–positron collider)

One important way to study particle interactions and to search for new particles is to accelerate an electron and a positron to high energies and then collide them, using a detector

23-7

to study what emerges. The energy to which they are accelerated is chosen to fit the question of interest; e.g. to study CP violation in b-quark decays the energy is chosen to maximize the production of b's in an appropriate way, while to produce new heavy particles the energy is made as large as possible. All uses of electron colliders require very large luminosity (intensity).

ELECTROMAGNETIC FORCE   See FORCE, ELECTROWEAK FORCE.

ELECTROWEAK FORCE   The descriptions of the electromagnetic and weak forces have been unified into a single description, the electroweak force. The electromagnetic and weak forces appear to be different because the W and Z bosons that mediate the weak force are massive, while the photon that mediates the electromagnetic force has no mass. Consequently it is easier to emit photons than W and Z bosons. The electroweak unified theoretical description treats all the bosons on an equal footing, and explains why they appear to be different, because of their masses.

ENERGY DENSITY   A density is an amount per unit volume.

ENERGY DENSITY OF UNIVERSE

$\sim 10^{-6}$ GeV cm^{-3}
Of galaxy, ~ 0.4 GeV cm^{-3}
Of starlight, $\sim \frac{1}{2}$ eV cm^{-3}
Of cosmic rays, ~ 1 eV cm^{-3}.

FAMILY   Quarks and leptons appear to come in three families, even though only one family appears to be needed to explain the world we see. The other families differ only in that they are heavier. We do not yet understand why there are three families, but we can fully describe their behavior. This is one of the main mysteries of particle physics. If one assumes M-theory in 11 dimensions at the Planck scale, before compactification, with a single 'particle' described by the Lie group E_8, then after renormalization group running to the electroweak scale it been resolved into the three families of quarks and leptons.

FERMION   Fermions are particles with half a unit of spin. They have different properties from particles with an integer unit of spin (bosons). Quarks and leptons, the matter particles, are fermions.

FEYNMAN DIAGRAMS   The rules of any quantum field theory can be formulated so that it is possible to draw a set of

(Feynman) diagrams representing the processes that can occur, and to assign a probability of occurrence to the process represented by each diagram or set of diagrams. The diagrams are very helpful guides to thinking about what processes can occur. The structure of the theory determines which diagrams are allowed.

FIELD

Every particle is the origin of a number of fields, one for each non-zero charge the particle carries. Interactions occur when one particle feels the field of another (and vice versa). There are electromagnetic fields, weak fields, and strong fields. Any particle with energy (remember, mass is a form of energy) sets up a gravitational field. In the Standard Model particles get mass by interacting with a Higgs field.

FINAL THEORY

The name used in this book for the theory sought by many particle physicists that not only includes the quantum Standard Model but also includes the theory of gravity, and explains why the final theory itself takes the form it does, and explains what quarks and other particles are. See THEORY OF EVERYTHING.

FORBIDDEN

Processes can be naïvely imagined that might occur, but should not occur according to the predictions of the Standard Model. Whether they occur is then a test of the Standard Model. If they occurred at the same rate as other processes the Standard Model would be wrong; if they occur at much smaller rates, or do not occur at all, they provide a clue as to how to extend the Standard Model. None of the processes forbidden by the Standard Model have been observed.

FORCE

All the phenomena we know of in nature can be described by five forces: the gravitational, weak, electric, magnetic, and strong forces. The electric and magnetic forces are unified into the electromagnetic force. Although the weak and electromagnetic forces appear different to us, they can be described as unified into one force (electroweak) in a more basic way; there is evidence that a similar unification of the electroweak force with the strong force also occurs. The attempt to unify all forces is an active research area. In particle physics, 'force' and 'interaction' mean essentially the same thing.

GAUGE BOSON	The strong, electromagnetic, and weak interactions are transmitted by the exchange of particles called gauge bosons (gluons, photons, and W's and Z's). The gauge bosons are the quanta of the strong, electromagnetic, and weak fields.
GAUGE KINETIC FUNCTION	A function arising from compactification. Once derived for a given compactification it can be used in calculations.
GAUGE THEORY	A quantum field theory where interactions occur between particles carrying charges, with strengths proportional to the sizes of the charges, and are transmitted by bosons that are quanta of the fields set up by the charges, is called a gauge theory. In a gauge theory, once any particle exists (such as an electron) that carries any kind of charge (electric or weak or strong or any yet to be found), the associated boson that transmits the force must exist (photons or W and Z bosons or gluons or any yet to be found), or the theory would not be consistent.
GAUGE SECTOR	A sector with only gauge fields, no fermions. For example, a pure SU(3) symmetry for a hidden sector.
GENERAL RELATIVITY	Einstein's theory of the gravitational interaction. Sometimes GR.
GLUINO	The hypothetical supersymmetric partner of the gluon, differing only in that the gluino has spin 1/2 while the gluon has spin 1, and the gluino is heavier.
GLUON	The particle that transmits the strong, or color, force; the quantum of the strong field.
GLUON JET	The color, or strong force is so strong that colored particles (quarks and gluons) hit or produced in a collision can only separate from other particles carrying color charge by binding to one of several hadrons. A quark or gluon appears in a detector as a jet of typically 5–15 hadrons.
GRAND UNIFICATION	The proposed unification of the weak, electromagnetic and strong forces into a single force. This unification, if it occurs, must happen in the sense that the forces act as a single one at very short distances, a million billion times smaller than distances that have been studied experimentally so far; the forces behave differently when they are studied at larger distances.
GRAVITATIONAL FORCE	See FORCE.

GRAVITON

The quantum of the gravitational field, that mediates the gravitational force.

HADRON

The properties of the color force and the rules of quantum theory allow certain combinations of quarks (and anti-quarks) and gluons to bind together to make a composite particle; all such particles are called hadrons. When mainly three quarks bind the resulting hadron is called a 'baryon'. When quark and anti-quark combine the result is called a 'meson', and when gluons combine it is called a 'glueball'. Hadrons have diameters of about 10^{-13} cm. The proton and neutron are the most familiar baryons. Pions are the lightest mesons so they are produced frequently in collisions. Kaons are the next lightest hadrons; their properties make them useful in many studies

HELIUM ABUNDANCE

As the Universe cooled after the Big Bang, it eventually reached a stage (about a minute after the beginning) when protons and neutrons formed, and then nuclei. Nuclei up to helium formed, but collisions were too soft for heavier nuclei to form. The theory of the Big Bang predicts the fraction of nuclei that are helium, and that fraction has been measured; the observed amount agrees very well with the predicted amount. This is one of the main reasons why it is generally believed that the Universe began in a hot Big Bang.

HIGGS BOSON

The Higgs boson is the quantum of the Higgs field. See HIGGS FIELD, HIGGS MECHANISM, HIGGS PHYSICS, and chapter 3.

HIGGS FIELD

See HIGGS MECHANISM. In the Standard Model particles (bosons and fermions) are thought to get mass by interacting with the Higgs field. The Higgs field and the way the particles interact with it must have very special properties for the masses to be included in the theory in a consistent way.

HIGGS MECHANISM

The Higgs mechanism is a special set of circumstances that must hold if bosons and fermions are to get masses from interacting with a Higgs field, even if the Higgs field exists. In the Standard Model these circumstances can be imposed, and in the supersymmetric Standard Model they can be derived. See chapter 3.

HIGGS PHYSICS	This is the combined physics that explains the origin of the Higgs field, the reason the Higgs mechanism applies, and the properties and study of the Higgs bosons.
HIGH ENERGY PHYSICS	Another name for particle physics, often used because much of particle physics (though not all) is based on experiments requiring high energy beams.
HIGGSINO	The superpartner of the Higgs boson.
HOT DARK MATTER	See DARK MATTER.
INFLATIONARY UNIVERSE	According to the inflationary universe theory, as the Universe expanded after the Big Bang, it went through a stage of very rapid expansion, called 'inflation', and then slowed down to the present rate. It began at Planck-scale size and ended up about the size of a baby.
INTENSITY	A measure of how often collisions occur at a collider. See LUMINOSITY.
INTERACTION	See FORCE.
JET	See GLUON JET.
KÄHLER POTENTIAL	A function arising from compactification, it can be thought of as the potential of the scalars. Once it is derived for a given compactification it can be used in calculations.
KAON	See HADRON.
KALUZA–KLEIN	Theodor Kaluza and Oskar Klein considered a 5D world, compactified to our 4D plus one curled up dimension, and showed the result could be interpreted as the 4D gravity plus a vector potential like electromagnetism. See Appendix B.
LAGRANGIAN	A Lagrangian is an equation that contains representations of all of the fundamental particles in the world, and specifies how they interact. Given the Lagrangian, the rules of quantum theory specify how to calculate the behavior of the particles, how to build up all of the composite systems they form, and all of the consequences of the basic theory
LHC (LARGE HADRON COLLIDER)	This is a CERN facility that collides protons in a circular ring 27 km in circumference, with beam energies of 6.5 TeV. The Higgs boson was discovered there in 2012. LHC will run again in 2022 (maybe earlier), then undergo a large luminosity increase and run again.
LEPTON	A class of particles defined by certain properties: leptons are fermions with spin 1/2 that do not carry color charge, and that have another

property called lepton number that is different for each family. The known leptons are the electron, the muon, the tau, and their associated neutrinos.

LIGHTEST SUPERPARTNER

The superpartner with the least mass may have several important roles. In particular, it may be the cold dark matter of the Universe, and its properties are crucial for identifying the events of superpartner production at colliders, since all of the heavier superpartners decay into the lightest one.

LINEAR ELECTRON COLLIDER

Particles traveling in a curved path continuously radiate photons that carry away some of the particles' energy. The fraction of energy radiated increases with the energy of the particle, and happens with greater probability for lighter particles than for heavier ones. For electrons this loss of energy was a large effect at the circular CERN LEP collider, and it would be worse at a higher energy collider. The next electron collider may be a linear one where the radiated energy loss is greatly decreased, modeled on the first linear collider, the SLC at SLAC. Two linear colliders are being designed, the ILC which would be built in Japan if it were funded, and CLIC which would be based at CERN if it were funded. More recently circular colliders of much larger diameter have been designed, to reduce the radiation. CERN, and China, are designing possible examples. They could be run first as electron–positron colliders in a tunnel of large diameter, and then proton–proton colliders.

LUMINOSITY

Any collider has two basic figures of merit, the maximum energy it can supply to the collisions, and how often it can cause collisions to occur. The number of events at a collider over some period of time is the product of two factors, the probability that if two particles actually collide something will happen, and the number of collisions. The latter is just a property of the collider, not of the physics that governs the collision. It is called the luminosity. It depends on features such as how many particles can be accelerated, how tightly bunches of them can be packed, and so forth. Loosely speaking we can refer to the luminosity as the intensity.

MANIFOLD	In mathematics, a manifold is a topological space that locally resembles Euclidean space near each point. More precisely, an n-dimensional manifold, or n-manifold for short, is a topological space with the property that each point has a neighborhood that is homeomorphic to the Euclidean space of dimension n (from Wikipedia).
M-THEORY	UV completion of relativistic quantum field theory.
MASS	Mass is an intrinsic property of any object, it measures how hard it is to make an object move. It can be thought of as weight, though the two are not quite the same (the mass of an object does not change, but if it were transported to a planet with a different mass its weight would change).
MAXWELL'S EQUATIONS	Electromagnetism, the unified theory of all electric and magnetic phenomena, is summarized in a set of equations, first written by Maxwell in the 1860s. When they are extended to include the effects of the quantum theory the theory of quantum electrodynamics (QED) is obtained. Physics students spend about a quarter of their time for two years learning how to solve Maxwell's equations, unless they plan to work in a subfield that relies heavily on Maxwell's equations, in which case they spend much more time studying them.
MATTER	It is useful to think of quarks and leptons as the basic particles that make up all the things around us, and the photons and gluons that bind them as quanta of the fields. We call the quarks and leptons matter particles. Sometimes 'matter particles' is used to mean fermions.
MATTER ASYMMETRY	Our Universe seems to be made of matter, but not antimatter, so there is an 'asymmetry'. Several ideas exist to explain how a universe could initially be symmetric, with equal numbers of protons and antiprotons, but evolve into our asymmetric one, with about a billion protons for every antiproton, today. This is an active research area. Sometimes called the baryon asymmetry. See chapter 13.
MEDIATE	The effects of interactions are transmitted from one particle to another by exchange of particles called bosons. The bosons are said to mediate the interaction or force.

MESON

See HADRON.

MICROWAVE BACKGROUND RADIATION

As the Universe expanded and cooled the original particles decayed or annihilated until only photons, neutrinos, protons, neutrons, and electrons that formed atoms remained. Today there is a cold gas of photons, about four hundred in each cubic centimeter of the Universe, called the microwave background radiation, because the wavelength of the photons is in the microwave part of the spectrum. The properties of this background radiation can tell us a great deal about the properties of the Universe and how it began, and are the subject of intense study.

MISSING ENERGY

When superpartners are produced at colliders, we expect each superpartner to decay into Standard Model particles plus the lightest superpartner, which interacts weakly and thus escapes the detector. The energy it carries off is expected to be one of the main signatures that tells us superpartners have been produced.

MODULI

Scalar fields that tell us how large Higgs fields are in the vacuum, and other scalar fields in general. The decay of the lightest modulus will occur shortly after inflation ends, and be the Big Bang.

MOLECULES

Although atoms are electrically neutral, the positive and negative charges are not on top of one another so there is some electric field outside an atom. Therefore atoms can attract each other and form molecules, which can get very large.

M-theory

M-theory has eleven dimensions (ten space). Compactified to four space–time dimensions on a 7D manifold of G_2 holonomy automatically gives a supersymmetric relativistic quantum field theory. M-theory is the main subject of this book. It seems to provide a rather complete description of our world, accounting for all observations. Compactified M-theory leads to the final theory.

MUON

A fundamental particle. A muon decays into an electron and neutrinos in about a millionth of a second. Muons are made in collisions at accelerators, and in decays of other particles produced at accelerators and in cosmic ray collisions. Very energetic muons can travel many meters before decaying, because of time dilation, so they can be very useful to detect at experiments.

NATIONAL LABORATORIES

Since much of the research in particle physics has to be done at large accelerators which are very expensive, the accelerators are built as national or international facilities at a few labs and used by all particle physicists. CERN is an international laboratory. Fermilab is a U.S. national laboratory. Particle physicists can carry out experiments at any laboratory.

NEUTRINO

A fundamental particle. There is one neutrino for each of the three families of particles.

NEUTRON

See HADRON. Free neutrons have a lifetime of about fifteen minutes; they decay into a proton and an electron and an anti-neutrino. When the neutrons are bound into nuclei (such as those in us) the decays are no longer possible because of subtle effects explained by quantum theory, so the neutrons in nuclei are as stable as protons.

NEWTON'S CONSTANT G

Newton's law of gravitation says that the gravitational force between any two bodies is proportional to the product of their masses and decreases as the square of the distance between them. G is measured by finding the force between two objects of known masses separated by a known distance. This statement is turned into an equation by inserting the constant G, so the force $F = Gmm'/r^2$. Since G is universal because all of the particles feel the gravitational force, G can be used to form quantities with dimensions, giving the Planck scale.

NEWTON'S LAWS

Newton formulated the law that describes the gravitational force (see NEWTON'S CONSTANT G), and three laws that describe motion. The first law says that every moving body moves in a straight line at constant speed unless a force acts on it, the second law says that the product of the mass of a body and its acceleration is equal to the force acting on it ($F = ma$), and the third law says that if one body applies a force on a second then the second applies an equal and oppositely directed force on the first.

NUCLEAR FORCE

Although protons and neutrons carry no strong or color charge, at tiny distances near a proton or neutron the cancellation of the strong field from its constituent quarks and gluons is incomplete, leaving a residual strong

	force that leaks outside the proton and neutron. This is the nuclear force that binds protons and neutrons into nuclei.
NUCLEUS	Although protons and neutrons are color-neutral composites of quarks and gluons, the quarks and gluons are not all at the same places so some of their color, or strong, fields exist outside the proton or neutron, giving an attractive force that binds protons and neutrons into nuclei. The attractive effects of this residual color force is offset by the electrical repulsion of the protons, so nuclei with too many protons cannot exist. It turns out that there are 92 stable or long-lived nuclei in nature. They are the nuclei of the atoms of the 92 chemical elements.
PARTICLE	'Particle' is used somewhat loosely, and includes not only the elementary quarks and leptons and bosons, but also the composite hadrons. It also includes any (currently hypothetical) new particles that might be discovered, such as the supersymmetric partners of the quarks and leptons and bosons.
PARTICLE PHYSICS	This is the field of physics that studies particles and tries to understand their behavior and properties. Sometimes a distinction is made between particle physics that studies quarks, leptons, gauge bosons and Higgs physics, and the study of hadron physics that aims to relate the properties of the hadrons to the theory of the color force. More broadly, the goals of particle physics are to understand not only the description of the particles and their interactions, but also why the laws of nature are what they are, and how the Universe arises from those laws.
PHOTINO	The supersymmetric partner of the photon.
PHOTON	The photon is the particle that makes up light. It transmits the electromagnetic force. It is the gauge boson of electromagnetism. Once electrons exist, quantum theory implies that photons must exist and have the properties they do.
PION	The lightest hadron, and therefore the one most often produced in collisions. See HADRON.
PLANCK'S CONSTANT h	Many things are quantized, such as the energy levels of atoms. h sets the scale of quantization —energy levels are separated by amounts

proportional to h, the amount of spin a particle can have is a multiple of h, etc. Planck originally found that the energy radiated by a heated body was emitted in quanta, rather than emission of any continuous amount being possible. The amount of energy was always an integer multiple of hf, where f is the frequency or color of the radiation; that is, hf of energy can be emitted, or $2hf$, or $3hf$ and so on, but not amounts in between. By separately measuring the frequency, Planck deduced the value of h, $h = 6.63 \times 10^{-34}$ Joule-seconds. See PLANCK SCALE.

PLANCK ENERGY, LENGTH, TIME   See PLANCK SCALE.

PLANCK SCALE   Planck scale refers to certain values of length, time, and energy or mass. To understand how these values originate, suppose you were trying to explain to an intelligent being in another galaxy how long humans typically lived. You couldn't use hours or years since those units are defined on earth (for example, by how long it happens to take our planet to go around its sun once), so a being in another galaxy wouldn't know what you meant. However, every scientist in the Universe knows the values of Planck's constant (h), the speed of light in vacuum (c), and the universal strength of the gravitational force (G). You could use those to form ratios that define a universal unit of time called the Planck time, and tell the being from another galaxy our typical lifetime in units of Planck times. Similar units for length and mass or energy can be defined. Max Planck realized this possibility and defined these units at the beginning of the 20th century. Since the Planck-scale units are the only universal ones, we expect the fundamental laws of nature to be simple in form when expressed in those units. See chapter 3.

POINT-LIKE   If matter is probed with projectiles that are large, and have energies that are less than what is needed to change the energy levels of an atom, then atoms will seem to be point-like objects. If the energy is increased eventually the projectile will penetrate the atom, but encounter the nucleus that will seem to be point-like. With higher energy the nucleus will appear to be made of point-like protons and neutrons. With still higher energies the protons and neutrons will be seen to be made of point-like quarks and gluons. As the energies of projectiles were increased still more, quarks and leptons might have been seen to be made of something still smaller, but that has not happened—they behave as point-like up to the highest energies they have been probed with, energies well beyond those for which we would have expected to find more constituents if history were to repeat itself once more. Further, the structure of the Standard Model theory suggests that quarks and leptons are the fundamental, point-like constituents of matter.

POSITRON
: The antiparticle of the electron.

PREDICT
: 'Predict' is used in the normal sense that a theory may predict some unanticipated or as yet unmeasured result. It is also used in another sense: a theory can be said to predict a result that is already known, because once the theory is written it gives a unique statement about that result. Sometimes an in-between situation holds, in that the theory predicts a result uniquely in principle, but the prediction depends on knowing some other quantity or requires very difficult calculations.

PROJECTILE
: To study particles and their interactions it is necessary to probe them with projectiles. The projectiles are other particles, electrons, photons, neutrinos and protons because these are small enough and can be given enough energy.

PROTON
: See HADRON.

PROTON DECAY
: If the Standard Model were the complete theory that described nature, protons would be stable, never decaying detectably. If the Standard Model is part of a more comprehensive theory that unifies quarks and leptons, then possibly protons are unstable, though with extremely long lifetimes. Some basic theories imply that protons decay, while others do not. Experiments that search for proton decay are very important, because if we knew it occurred (and what the proton decayed into) it would provide valuable information about how to extend the Standard Model.

QUANTA
: Each particle is surrounded by a field for each of the kinds of charges it carries, such as an electromagnetic field if it has electrical charge. In the quantum theory the field is described as made up of particles that are the quanta of the field. More loosely, the smallest amount of something that can exist.

QUANTUM FIELD THEORY

When interactions among particles are described as transmitted by exchange of bosons the methods of quantum field theory are used.

QUANTUM THEORY

The quantum theory provides the rules to calculate how matter behaves. Once scientists specify what system they want to describe, and what the interactions among the particles of the system are, then the equations of the quantum theory are solved to learn the properties of the system.

QUARK

A fundamental particle. Quarks are very much like electrons, but also carry strong charge and thus have another interaction, one that can bind them into protons and neutrons. There are six quarks, called up (u), down (d), charm (c), strange (s), top (t), and bottom (b).

QUARK JET

Because quarks must end up in hadrons, quarks that are produced in collisions actually appear in detectors as a narrow jet of hadrons, mostly pions. See GLUON JET.

RADIOACTIVE DECAY

Some nuclei are unstable, but live long enough to exist as matter until they decay. When they decay they can emit several particles, photons, electrons, positrons, neutrinos, neutrons, and even helium nuclei. For historical reasons such decays were called 'radioactive decays'. Sometimes scientists use the emitted particles as tools to do experiments.

REDUCTIONIST

One way to study the natural universe is to study very detailed aspects of nature, to take things apart and see what they are made of, and to focus on small steps. This approach is called 'reductionist'. It has been a powerful success, letting us build up the remarkably complete description of nature we now have. Whenever possible, scientists have tried to unify subfields as they became understood. Recently in particle physics the trend toward unification has been increasingly successful. For physicists, reductionism includes the associated unification. See UNIFICATION.

RELATIVISTIC, RELATIVISTIC INVARIANCE

Whenever particles can move at speeds near the speed of light, and whenever fields are involved, the description of nature must satisfy the requirements of Einstein's 'special relativity' theory.

RULES

In order to have a complete understanding of nature, it is necessary to know the particles, the forces that determine the interactions of the particles, and the rules for calculating how the particles behave. For the motion of objects normally on earth or in the sky, the rule to use to calculate the behavior of particles is Newton's second law, $F = ma$. When atomic distances or smaller are involved, the Schrödinger equation of quantum theory replaces Newton's second law. In particle physics additional relativistic requirements are added to make the complete set of rules, quantum theory and Einstein's special relativity. See chapter 1.

SCHRÖDINGER EQUATION

The equation from quantum theory that tells how to calculate the effects of the forces on the particles. It is the quantum theory equivalent of Newton's second law.

SCIENCE

Science can be defined as a self-correcting way to get knowledge about the natural universe, plus the body of knowledge obtained that way. It is both a method and the resulting understanding and knowledge. The method requires making models to explain phenomena, testing them experimentally, and revising them until they work. The goal of science is understanding. Once part of the natural world is understood, it may be possible to develop applications of the new knowledge. The process of developing such applications is properly called technology, not science. Although scientific knowledge may, and usually does, lead to technology, science is not necessary for technology, and technological developments have led to new science as much as the opposite. Before the time of Galileo many technological developments occurred that had no scientific connection. Since the time of Maxwell and his writing of the electromagnetic theory almost all technological developments have depended on earlier science. In recent years the words 'science' and 'technology' have been frequently misused, as if they were interchangeable. Because science and technology are really different, it is better to carefully distinguish between them.

SELECTRON

The supersymmetric partner of the electron.

SIGNATURE

A new particle will have some characteristic behavior in a detector that allows it to be recognized. Particles that decay into others do so in a unique way that is different for every particle. Knowing the properties of the particle allows us to calculate how it will decay. The features that allow a new particle to be identified in a detector are called its signature.

SLEPTON

The supersymmetric partner of any of the leptons.

SMATTER

The superpartners of the Standard Model particles. This book argues that the experimental discovery of smatter will provide us with information that will be essential for gaining insights into the ultimate laws of nature, the final theory.

SOLAR NEUTRINOS

The reactions that fuel the sun lead to the emission of photons that reach the earth as sunlight, and of neutrinos that we do not see with our eyes, but which can be detected in special neutrino detectors. At present there is great interest in these neutrinos because the number being detected is fewer than expected, and this may be a signal that neutrinos have mass, in which case we could account for the fewer number detected. If they have mass, the experiments to detect them will allow the value of their mass to be measured.

SPECIAL RELATIVITY

The constraints of special relativity are two conditions that Einstein pointed out should be satisfied by any acceptable physical theory. Somewhat oversimplified, these conditions are first that light moves at the same speed in vacuum regardless of how it is emitted, and second that scientists working in different labs moving with different relative speeds should formulate the same natural laws. The constraints imposed by these conditions have surprising implications for the structure of acceptable theories. For example, the Schrödinger equation of quantum theory does not satisfy these conditions. But when it was generalized by Dirac to do so, the resulting equation led to the prediction of antiparticles, which need not have existed from the point of view of quantum theory alone.

SPECTRA

Atoms can exist in a number of discreet energy levels. They emit or absorb photons when they make transitions from one level to another. The energies of the photons emitted or absorbed by one atom are different from those of all other atoms. The photon energies are directly related to their frequencies, which set their colors in the spectrum, so by observing the colors of the photons it is possible to determine which atoms are being observed. This can be done in a laboratory, and it can also be done with the light reaching us from stars, near or distant, which allows us to identify the atoms that stars are made of. Only the same 92 elements we find on earth are seen throughout the Universe.

SPEED OF LIGHT — Light and all other massless particles travel in vacuum with a speed, usually labeled c, whose value is about three hundred million meters a second. Special relativity implies that no particle or signal can move faster than the speed of light, and that photons always have this speed regardless of the speed of their source.

SPIN — Spin is a property that all particles have. It is as if particles were always spinning at a fixed rate (which could be zero), which can be different according to the type of particle. It is not quite right to think of them actually spinning because the particles do not have to have spatial extension to have spin; calling this property 'spin' is an analogy. The amount of spin is required by the quantum theory to come in definite amounts; if the unit is chosen to be Planck's constant, h divided by 2π, then particles can have zero spin, half a unit of spin, one unit of spin, etc.

SPONTANEOUS SYMMETRY BREAKING — Often the equations of a theory may have certain symmetries, though their solutions may not; the symmetries are hidden, or broken. For example, the equations may describe several particles in identical ways, so the equations are unchanged if the particles are interchanged, but the solutions may give the particles different properties (this is illustrated in a simple example in chapter 1). This phenomenon is called spontaneous symmetry breaking.

SQUARK — The supersymmetric partner of any of the quarks.

STABLE PARTICLE — Some particles do not decay into others; they are called stable. See DECAY.

STANDARD MODEL — The very successful theory of quarks and leptons and their interactions that is described in this book is called the 'Standard Model' by particle physicists. The name arose historically as the theory developed, and then was difficult to change because it was widely used. The Standard Model is the most complete mathematical theory of the natural world ever developed, and is well tested experimentally.

STRING PHENOMENOLOGY   String phenomenology is starting with a high scale 10D or 11D theory, compactifying, stabilizing moduli, and explaining or predicting observables.

STRING THEORY   String theory is a theory that aims to unify all of the forces and particles of nature and explain why they are as they are. In string theory there is only one force (gravity), in nine or ten space dimensions, but when looked at from our four-dimensional world the extra dimensions imply the other forces we observe. Particles are strings that vibrate in different ways to account for their various properties. Whereas field theory with point-like particles is too singular and leads to divergences, string theory by dealing with extended objects is a finite theory. String theories appear to allow the construction of a quantum theory of gravity.

STRONG FORCE   See FORCE.

STRUCTURE   Objects have structure if they have parts—if they are made of other things. Whether objects have structure can be learned from experiments that probe them with projectiles. Over the past century each stage of matter was found as it became possible to search for smaller things that turned out to have structure. Quarks and leptons appear not to have structure, so perhaps the search for the basic constituents has finally ended. There are also theoretical arguments that quarks and leptons are the basic constituents.

SUBATOMIC PARTICLE   Any particle that is contained in an atom, or any particle that can be created in collisions of such particles, is loosely called 'subatomic,' whether it is composite like a proton, or elementary like a quark or electron.

SUPERPARTNER   If the theory that describes nature has a symmetry called supersymmetry, then every normal particle (the ones we know) has associated with it a partner that differs only in its spin and, and possibly its mass.

SUPERPOTENTIAL   A function arising from compactifications, somewhat analogous to a Lagrangian.

SUPERSPACE | Supersymmetry can be formulated in several ways. One is to imagine associating another coordinate having special properties with each of our normal space–time coordinates, giving a kind of space called superspace. Writing theories in superspace makes them supersymmetric. This way of constructing supersymmetric theories is harder to picture than associating superpartners with each Standard Model particle, but leads to the same results and sometimes facilitates deriving mathematical properties of the theories.

SUPERSTRING | String theories are expected to be supersymmetric, and are often called superstring theories.

SUPERSYMMETRY | A hypothetical symmetry that describes nature, which says that even though fermions and bosons seem to us to be very different in their properties and their roles, in the theory itself they appear in a symmetric way. If supersymmetry is indeed realized in nature, then every particle has a superpartner.

SUSY | This is a common abbreviation for supersymmetry

TECHNOLOGY | See SCIENCE.

THEORY | The word 'theory' is usually used precisely in physics. Theories are not conjectures, but sets of equations whose solutions describe physical systems and their behavior.

THEORY OF EVERYTHING | A 'theory of everything' would not only describe how things work, it would explain why things are the way they are, the black hole information problem, etc. The name 'theory of everything' is unfortunate in one way because it does not tell how to deduce the behavior and emergence of complex systems from a knowledge of their components. It's much more ambitious than the 'final theory'.

TRANSMIT | See MEDIATE.

THERMAL (COSMOLOGICAL) HISTORY | The conventional history. After the Big Bang the Universe cools. From compactified theories the history is dominated by moduli stabilizing and decaying.

TREE LEVEL, TREE DIAGRAM | Feynman diagrams with no internal structure.

UV COMPLETE | The ultraviolet is the higher energy end of the spectrum, so 'UV complete' means having a quantum gravity theory included.

UNCERTAINTY PRINCIPLE | The uncertainty principle is a consequence of quantum theory. It implies that a pair of observables cannot both be measured simultaneously to arbitrary accuracy. It can often be used to understand quantum theory results in a simple way.

UNIFICATION

Scientists have sought for centuries to unify the descriptions of apparently different phenomena by showing they were due to the same underlying natural laws, and that complex levels of matter were made of simpler levels. This unification process is a subject of very active research for the forces of nature today. The possible unification of the strong, electromagnetic, and weak forces is called a 'grand unification'. There is a continuing effort to unify these forces with gravity. String theories seem to do that successfully.

UNSTABLE PARTICLE

See DECAY.

VACUUM

Any physical system will settle into the lowest energy state it can, which we call its vacuum state in particle physics. For most systems that is the state where the fields making up the system are zero, but theorists hypothesize that for systems containing Higgs fields the lowest energy occurs when the Higgs field takes on a constant value different from zero. The value of the Higgs field in that system is called its 'vacuum expectation value'.

VACUUM EXPECTATION VALUE

See VACUUM.

VOLUME MODULUS

The combination of moduli that gives the volume of the Planck-scale region in the seven-dimensional space.

WEAK FORCE

See FORCE. The weak force is described in chapter 4.

WEAK CHARGE

See CHARGE.

WINO

The supersymmetric partner of the W boson.

YANG–MILLS FORCE

A special kind of field theory, with charges the particles can carry, and which determine the strength of the force. The Standard Model forces are Yang–Mills forces.

ZINO

The supersymmetric partner of the Z boson.

CPSIA information can be obtained
at www.ICGtesting.com
Printed in the USA
BVHW061925171021
619145BV00003B/69